Robert Boyle, Carl Barus, Émile Hilaire Amagat

The Laws of Gases

Memoirs

Robert Boyle, Carl Barus, Émile Hilaire Amagat

The Laws of Gases
Memoirs

ISBN/EAN: 9783744666411

Printed in Europe, USA, Canada, Australia, Japan

Cover: Foto ©Suzi / pixelio.de

More available books at **www.hansebooks.com**

HARPER'S SCIENTIFIC MEMOIRS

EDITED BY

J. S. AMES, Ph.D.

PROFESSOR OF PHYSICS IN JOHNS HOPKINS UNIVERSITY

V.

THE LAWS OF GASES

THE LAWS OF GASES

MEMOIRS BY ROBERT BOYLE AND
E. H. AMAGAT

TRANSLATED AND EDITED BY

CARL BARUS
PROFESSOR OF PHYSICS IN BROWN UNIVERSITY

NEW YORK AND LONDON
HARPER & BROTHERS PUBLISHERS
1899

HARPER'S SCIENTIFIC MEMOIRS.

EDITED BY

J. S. AMES, Ph.D.,

PROFESSOR OF PHYSICS IN JOHNS HOPKINS UNIVERSITY.

NOW READY:

THE FREE EXPANSION OF GASES. Memoirs by Gay - Lussac, Joule, and Joule and Thomson. Editor, Prof. J. S. Ames, Ph.D., Johns Hopkins University. 75 cents.

PRISMATIC AND DIFFRACTION SPECTRA. Memoirs by Joseph von Fraunhofer. Editor, Prof. J. S. Ames, Ph.D., Johns Hopkins University. 60 cents.

RÖNTGEN RAYS. Memoirs by Röntgen, Stokes, and J. J. Thomson. Editor, Prof. George F. Barker, University of Pennsylvania. 60 cents.

THE MODERN THEORY OF SOLUTION. Memoirs by Pfeffer, Van't Hoff, Arrhenius, and Raoult. Editor, Dr. H. C. Jones, Johns Hopkins University. $1 00.

THE LAWS OF GASES. Memoirs by Boyle and Amagat. Editor, Prof. Carl Barus, Brown University.

IN PREPARATION:

THE SECOND LAW OF THERMODYNAMICS. Memoirs by Carnot, Clausius, and Thomson. Editor, Prof. W. F. Magie, Princeton University.

THE FUNDAMENTAL LAWS OF ELECTROLYTIC CONDUCTION. Memoirs by Faraday, Hittorf, and Kohlrausch. Editor, Dr. H. M. Goodwin, Massachusetts Institute of Technology.

THE EFFECTS OF A MAGNETIC FIELD ON RADIATION. Memoirs by Faraday, Kerr, and Zeeman. Editor, Dr. E. P. Lewis, University of California.

WAVE-THEORY OF LIGHT. Memoirs by Huygens, Young, and Fresnel. Editor, Prof. Henry Crew, Northwestern University.

NEWTON'S LAW OF GRAVITATION. Editor, Prof. A. S. Mackenzie, Bryn Mawr College.

NEW YORK AND LONDON:

HARPER & BROTHERS, PUBLISHERS.

PREFACE

OF course anybody may read the famous *Memoirs* of Amagat in the original; but everybody cannot so easily get these papers permanently into his possession. I believe, therefore, with the present translations to have scored a point in the interest of accessibility, and thus to have materially contributed to the advancement of science.

C. B.

BROWN UNIVERSITY, Providence, R. I.,
March, 1899.

83054

GENERAL CONTENTS

A DEFENCE OF
THE DOCTRINE TOUCHING THE SPRING
AND WEIGHT OF THE AIR

Proposed by Mr. R. Boyle in his *New Physico-Mechanical Experiments,* against the Objections of Franciscus Linus, wherewith the Objector's Funicular Hypothesis is also Examined

London, 1662

A DEFENCE OF
THE DOCTRINE TOUCHING THE SPRING
AND WEIGHT OF THE AIR*

By ROBERT BOYLE

PART II., CHAPTER V.

TWO NEW EXPERIMENTS TOUCHING THE MEASURE OF THE FORCE OF THE SPRING OF AIR COMPRESSED AND DILATED

. . . .

We took then a long glass-tube, which, by a dexterous hand and the help of a lamp, was in such a manner crooked at the bottom, that the part turned up was almost parallel to the rest of the tube, and the orifice of this shorter leg of the siphon (if I may so call the whole instrument) being hermetically sealed, the length of it was divided into inches (each of which was subdivided into eight parts) by a streight list of paper, which containing those divisions, was carefully pasted all along it. Then putting in as much quicksilver as served to fill the arch or bended part of the siphon, that the mercury standing in a level might reach in the one leg to the bottom of the divided paper, and just to the same height or horizontal line in the other ; we took care, by frequently inclining the tube, so that the air might freely pass from one leg into the other by the sides of the mercury (we took, I say, care) that the air at last included in the shorter cylinder should be of the same laxity with the rest of the air about it.   This done, we began to pour quicksilver into the longer leg of the siphon, which by its weight pressing up that in the shorter leg, did by degrees streighten

---

*Selected from Boyle's *New Physico - Mechanical Experiments*, London, 1662.

3

the included air : and continuing this pouring in of quicksilver
till the air in the shorter leg was by condensation reduced to
take up but half the space it possessed (I say, possessed, not
filled) before ; we cast our eyes upon the longer leg of the glass,
on which was likewise pasted a list of paper carefully divided
into inches and parts, and we observed, not without delight
and satisfaction, that the quicksilver in that longer part of the
tube was 29 inches higher than the other. Now that this
observation does both very well agree with and confirm our
hypothesis, will be easily discerned by him that takes notice
what we teach ; and Monsieur Paschal and our English friend's
experiments prove, that the greater the weight is that leans
upon the air, the more forcible is its endeavour of dilatation,
and consequently its power of resistance (as other springs are
stronger when bent by greater weights). For this being considered, it will appear to agree rarely-well with the hypothesis,
that as according to it the air in that degree of density and
correspondent measure of resistance, to which the weight of
the incumbent atmosphere had brought it, was able to counterbalance and resist the pressure of a mercurial cylinder of about
29 inches, as we are taught by the Torricellian experiment ;
so here the same air being brought to a degree of density about
twice as great as that it had before, obtains a spring twice as
strong as formerly. As may appear by its being able to sustain
or resist a cylinder of 29 inches in the longer tube, together
with the weight of the atmospherical cylinder, that leaned upon
those 29 inches of mercury ; and, as we just now inferred from
the Torricellian experiment, was equivalent to them.

We were hindered from prosecuting the trial at that time
by the casual breaking of the tube. But because an accurate
experiment of this nature would be of great importance to the
doctrine of the spring of the air, and has not yet been made
(that I know) by any man ; and because also it is more uneasy
to be made than one would think, in regard of the difficulty as
well of procuring crooked tubes fit for the purpose, as of making
a just estimate of the true place of the protuberant mercury's
surface ; I suppose it will not be unwelcome to the reader to be
informed, that after some other trials, one of which we made
in a tube whose longer leg was perpendicular, and the other,
that contained the air, parallel to the horizon, we at last pro-
cured a tube of the figure expressed in the scheme; which

4

tube, though of a pretty bigness, was so long, that the cylinder, whereof the shorter leg of it consisted, admitted a list of paper, which had before been divided into 12 inches and their quarters, and the longer leg admitted another list of paper of divers feet in length, and divided after the same manner. Then quicksilver being poured in to fill up the bended part of the glass, that the surface of it in either leg might rest in the same horizontal line, as we lately taught, there was more and more quicksilver poured into the longer tube ; and notice being watchfully taken how far the mercury was risen in that longer tube, when it appeared to have ascended to any of the divisions in the shorter tube, the several observations that were thus successively made, and as they were made set down, afforded us the ensuing table :

A TABLE OF THE CONDENSATION OF THE AIR

A	A	B	C	D	E	
48	12	00		29 2/16	29 2/16	*AA.* The number of equal spaces in the shorter leg, that contained the same parcel of air diversely extended.
46	11½	01 7/16		30 9/16	33 6/16	
44	11·	02 13/16		31 15/16	31 12/16	
42	10½	04 6/16		33 8/16	33¼	
40	10	06 3/16		35 5/16	35	
38	9½	07 14/16		37	36 15/16	*B.* The height of the mercurial cylinder in the longer leg, that compressed the air into those dimensions.
36	9	10 2/16		39 5/16	38 7/8	
34	8½	12 8/16		41 10/16	41 2/17	
32	8	15 1/16		44 3/16	43 11/16	
30	7½	17 15/16	Added to 22 2/8 makes	47 1/16	46 3/4	
28	7	21 3/16		50 5/16	50	*C.* The height of the mercurial cylinder, that counterbalanced the pressure of the atmosphere.
26	6½	25 3/16		54 6/16	53 10/16	
24	6	29 11/16		58 13/16	58 2/8	
23	5¾	32 3/16		61 5/16	60 18/23	
22	5½	34 15/16		64 1/16	63 6/17	
21	5¼	37 15/16		67 1/16	66¼	*D.* The aggregate of the two last columns, *B* and *C*, exhibiting the pressure sustained by the included air.
20	5	41 9/16		70 11/16	70	
19	4¾	45		74 2/16	73 11/16	
18	4½	48 12/16		77 14/16	77 2/8	
17	4¼	53 11/16		82 12/16	82 4/17	
16	4	58 2/16		87 14/16	87 3/8	*E.* What that pressure should be according to the hypothesis, that supposes the pressures and expansions to be in reciprocal proportion.
15	3¾	63 15/16		93 1/16	93 1/8	
14	3½	71 5/16		100 7/16	99 4/8	
13	3¼	78 11/16		107 13/16	107 1/13	
12	3	88 7/16		117 9/16	116 4/8	

5

For the better understanding of this experiment, it may not be amiss to take notice of the following particulars :

1. That the tube being so tall that we could not conveniently make use of it in a chamber, we were fain to use it on a pair of stairs, which were yet very lightsome, the tube being for preservation's sake by strings so suspended that it did scarce . touch the box presently to be mentioned.

2. The lower and crooked part of the pipe was placed in a square wooden box, of a good largeness and depth, to prevent the loss of the quicksilver, that might fall aside in the transfu-'sion from the vessel into the pipe, and to receive the whole quicksilver in case the tube should break.

3. That we were two to make the observation together, the one to take notice at the bottom, how the quicksilver rose in the shorter cylinder, and the other to pour in at the top of the longer ; it being very hard and troublesome for one man alone to do both accurately.

4. That the quicksilver was poured in but by little and little, according to the direction of him that observed below ; it being far easier to pour in more than to take out any, in case too much at once had been poured in.

5. That at the beginning of the operation, that we might the more truly discern where the quicksilver rested from time to time, we made use of a small looking-glass held in a convenient posture to reflect to the eye what we desired to discern.

6. That when the air was so comprest as to be crouded into less than a quarter of the space it possessed before, we tried whether the cold of a linen cloth dipped in water would then condense it. And it sometimes seemed a little to shrink, but not so manifestly that we dare build anything upon it. We then tried likewise, whether heat would, notwithstanding so forcible a compressure, dilate it ; and approaching the flame of a candle to that part where the air was pent up, the heat had a more sensible operation than the cold had before ; so that we scarce doubted, but that the expansion of the air would, notwithstanding the weight that comprest it, have been made conspicuous, if the fear of unseasonably breaking the glass had not kept us from increasing the heat.

Now although we deny not, but that in our table some particulars do not so exactly answer to what our formerly mentioned hypothesis might perchance invite the reader to expect ;

yet the variations are not so considerable, but that they may probably enough be ascribed to some such want of exactness as in such nice experiments is scarce avoidable. But for all that, till further trial hath more clearly informed me, I shall not venture to determine whether or no the intimated theory will hold universally and precisely, either in condensation of air or rarefaction : all that I shall now urge being, that however the trial already made sufficiently proves the main thing, for which I here allege it ; since by it, it is evident, that as common air, when reduced to half its wonted extent, obtained near about twice as forcible a spring as it had before ; so this thus comprest air being further thrust into half this narrow room, obtained thereby a spring about as strong again as that it last had, and consequently four times as strong as that of the common air. And there is no cause to doubt, that if we had been here furnished with a greater quantity of quicksilver and a very strong tube, we might, by a further compression of the included air, have made it counterbalance the pressure of a far taller and heavier cylinder of mercury. For no man perhaps yet knows how near to an infinite compression the air may be capable of, if the compressing force be competently increased.

. . . .

And to let you see, that we did not (as a little above) inconsiderately mention the weight of the incumbent atmospherical cylinder as a part of the weight resisted by the imprisoned air, we will here annex, that we took care; when the mercurial cylinder in the longer leg of the pipe was about an hundred inches high, to cause one to suck at the open orifice ; whereupon (as we expected) the mercury in the tube did notably ascend. . . . And therefore we shall render this reason of it that the pressure of the incumbent air being in part taken off by its expanding itself into the sucker's dilated chest, the imprisoned air was thereby enabled to dilate itself manifestly, and repel the mercury, that comprest it, till there was an equality of force betwixt the strong spring of that comprest air on the one part, and the tall mercurial cylinder, together with the contiguous dilated air, on the other part.

Now, if to what we have thus delivered concerning the compression of the air, we add some observations concerning its spontaneous expansion, it will the better appear, how much the phænomena of these mercurial experiments depend upon the differing measures of strength to be met with in the air's spring, according to its various degrees of compression and laxity.

.   .   .   .

A TABLE OF THE RAREFACTION OF THE AIR

	A	B	C	D	E
A. The number of equal spaces at the top of the tube, that contained the same parcel of air.	1	$00\frac{2}{3}$		$29\frac{3}{4}$	$29\frac{3}{4}$
	$1\frac{1}{2}$	$10\frac{4}{5}$		$19\frac{1}{8}$	$19\frac{4}{5}$
	2	$15\frac{3}{8}$		$14\frac{3}{8}$	$14\frac{1}{8}$
	3	$20\frac{2}{3}$		$9\frac{4}{5}$	$9\frac{15}{16}$
B. The height of the mercurial cylinder, that together with the spring of the included air counterbalanced the pressure of the atmosphere.	4	$22\frac{5}{8}$		$7\frac{1}{8}$	$7\frac{7}{16}$
	5	$24\frac{1}{4}$		$5\frac{5}{8}$	$5\frac{19}{20}$
	6	$24\frac{7}{8}$		$4\frac{7}{8}$	$4\frac{27}{40}$
	7	$25\frac{1}{8}$		$4\frac{2}{3}$	$4\frac{1}{4}$
	8	$26\frac{2}{3}$		$3\frac{4}{5}$	$3\frac{33}{40}$
	9	$26\frac{3}{4}$		$3\frac{3}{8}$	$3\frac{11}{40}$
C. The pressure of the atmosphere.	10	$26\frac{5}{8}$	Subtracted from $29\frac{1}{4}$ leaves	$3\frac{0}{8}$	$2\frac{19}{20}$
	12	$27\frac{1}{8}$		$2\frac{5}{8}$	$2\frac{23}{28}$
	14	$27\frac{4}{5}$		$2\frac{2}{3}$	$2\frac{1}{8}$
D. The complement of B to C, exhibiting the pressure sustained by the included air.	16	$27\frac{7}{8}$		$2\frac{0}{8}$	$1\frac{15}{16}$
	18	$27\frac{7}{8}$		$1\frac{7}{8}$	$1\frac{47}{48}$
	20	$28\frac{2}{3}$		$1\frac{5}{8}$	$1\frac{9}{80}$
	24	$28\frac{2}{3}$		$1\frac{1}{4}$	$1\frac{33}{48}$
E. What that pressure should be, according to the hypothesis.	28	$28\frac{3}{4}$		$1\frac{3}{8}$	$1\frac{1}{16}$
	32	$28\frac{1}{4}$		$1\frac{1}{8}$	$0\frac{11}{128}$

To make the experiment of the debilitated force of expanded air the plainer, it will not be amiss to note some particulars, especially touching the manner of making the trial; which (for the reasons lately mentioned) we have made on a lightsome pair of stairs, and with a box also lined with paper to receive the mercury that might be spilt. And in regard it would require a vast and in few places procurable quantity of quicksilver, to imploy vessels of such kind as are ordinary in the Torricellian experiment, we made use of a glass-tube of about six feet long; for that being hermetically sealed at one end, served our turn as well as if we could have made the experiment in a tub or pond of seventy inches deep.

Secondly, We also provided a slender glass-pipe of about the bigness of a swan's quill, and open at both ends; all along which was pasted a narrow list of paper, divided into inches and half quarters.

Thirdly, This slender pipe being thrust down into the greater tube almost filled with quicksilver, the glass helped to make it swell to the top of the tube; and the quicksilver getting in at the lower orifice of the pipe, filled it up until the mercury included in that was near about a level with the surface of the surrounding mercury in the tube.

Fourthly, There being, as near as we could guess, little more than an inch of the slender pipe left above the surface of the restagnant mercury, and consequently unfilled therewith, the prominent orifice was carefully closed with sealing wax melted; after which the pipe was let alone for a while, that the air, dilated a little by the heat of the wax, might, upon refrigeration, be reduced to its wonted density. And then we observed by the help of the above mentioned list of paper, whether we had not included somewhat more or somewhat less than an inch of air; and in either case we were fain to rectify the error by a small hole made (with a heated pin) in the wax, and afterwards closed up again.

Fifthly, Having thus included a just inch of air, we lifted up the slender pipe by degrees, till the air was dilated to an inch, an inch and an half, two inches, &c., and observed in inches and eighths the length of the mercurial cylinder, which at each degree of the air's expansion was impelled above the surface of the restagnant mercury in the tube.

Sixthly, The observations being ended, we presently made the Torricellian experiment with the above-mentioned great tube of six feet long, that we might know the height of the mercurial cylinder, for that particular day and hour; which height we found to be 29¾ inches.

Seventhly, Our observations made after this manner furnished us with the preceding table, in which there would not probably have been found the difference here set down betwixt the force of the air, when expanded to double its former dimensions, and what that force should have been precisely according to the theory, but that the included inch of air received some little accession during the trial; which this newly mentioned difference making us suspect, we found by replunging the pipe into the quicksilver, that the included air had gained about

9

half an eighth, which we guessed to have come from some little aerial bubbles in the quicksilver, contained in the pipe (so easy is it in such nice experiments to miss of exactness).

. . . .

The Honorable ROBERT BOYLE was born in Ireland, County Cork, January 25, 1626, and died in London, December 30, 1691. He made his home in Oxford from 1654 to 1668, when he moved to London. It was while living at Oxford that he invented the air-pump, which was perfected for him in 1658 or 1659 by his assistant in chemistry, Robert Hooke, who was afterwards so famous. In 1663, at the incorporation of the Royal Society by King Charles II., Boyle was appointed by the charter one of the council, as he had been one of the persons to whom the society owed its origin. He was elected president of the society in November, 1680 ; but on account of "the obligation to take the test and oaths," he felt obliged to decline the honor.

It was in 1660 that he published his first experiments, *On the Spring of the Air*, and in 1662 that he announced the law of gases that bears his name. These experiments were repeated later, by Mariotte in France, and were published by him in 1676.

Boyle published a great many papers in the *Philosophical Transactions*, and was actively engaged in scientific work up to the last years of his life. It is worthy of note that nearly all the common lecture - experiments in hydrostatics and pneumatics are due to Boyle.

ON THE COMPRESSIBILITY OF GASES AT HIGH PRESSURES

BY

E. H. AMAGAT

(*Annales de Chimie et de Physique*, 5ᵉ série, t. xxii., pp. 353–398, 1881)

CONTENTS

ON THE COMPRESSIBILITY OF GASES AT HIGH PRESSURES

By E. H. AMAGAT

INTRODUCTORY

In 1869, when I communicated the results of my first experiments relative to the effect of temperature on the compressibility of a gas, no direct experiments had been made on this subject. It was customary merely to quote Regnault's inferences touching the behavior of carbon dioxide at 100° C. At this temperature, since the density of the gas is sensibly independent of pressure for pressures in the neighborhood of 1 atmosphere, one may conclude that under the conditions stated the gas appreciably follows Mariotte's law. Nevertheless, the inference of Regnault is in need of slight modification. My results proved that at 100°, carbon dioxide, although diverging less from the law than at ordinary temperatures, is quite sensibly divergent; and M. Blaserna showed[*] a short time afterwards that the discrepancy arose out of an error which slipped into the numerical calculations of Regnault. Indeed, M. Blaserna arrived at an analogous conclusion himself in a purely theoretical paper published in the *Annales de Chimie et de Physique* in 1865 (t. v.).

In 1872 I made a complete publication[†] of my researches on this subject, having studied ammonia as far as 100°, sulphur dioxide and carbon dioxide as far as 250°, and air and hydrogen up to 320°.

I then believed that at temperatures sufficiently low all gases

---

[*] *Comptes Rendus des Séances de l'Académie des Sciences*, t. xlix., 1869.
[†] *Annales de Chimie et de Physique*, t. xxix., 1873.

behave like carbon dioxide; but that as temperature increases
gases begin more and more to follow the law of Mariotte, finally
to depart from it indefinitely in a contrary sense as does hydrogen at ordinary atmospheric temperatures. It follows naturally that nitrogen, which, as I showed, obeys Mariotte's law
as far as 100° and between 1 and 2 atmospheres, presents a negative discrepancy below that temperature; and that hydrogen
more and more highly heated continues to present an invariably negative divergence of increasing value. Experiment,
however, has shown me that at 200° the divergence of air always sensibly zero does not from the shape of the curves tend
to become negative, and that even for the case of hydrogen
the discrepancy tends rather to vanish than to increase in the
negative direction or to remain negative. Naturally I then inferred that the effect of temperature was an approach of both
gases towards the law, the compressibility of the first being diminished and that of the other increased.

It will be seen in the course of the present paper that this
conclusion is correct relative to hydrogen, but that for air the
interpretation of the results will have to be changed. It will
also be seen that it must be impossible to arrive at a knowledge
of the general laws for gases as long as the investigations are
limited to an interval of pressure within which the trend of
the discrepancies hardly even begins to appear; that these laws
will not be manifest in their entirety until the experiments are
pushed forward throughout several hundred atmospheres; that,
finally, all that has hitherto been known about the compressibility of gases would not have enabled any one to even suspect
the laws in their true characters, such as I shall establish them
below.

I shall not enter into any details as to the formulas by aid
of which different physicists have sought to express the effect
of temperature on the compressibility of gases. These formulas, as a rule purely empyric, are not applicable except within
very narrow pressure-limits; moreover, I shall have occasion to
speak of them in the near future in a paper referring specially
to low pressures.

During the course of my experiments I was able to show that,
even after making allowance for atomic volume, the shortcom-
ings of Mariotte's law cannot be explained fully by postulat-
ing an internal pressure tending to move the molecules nearer

together, because this pressure would be different for the same mean distance of the molecules at different temperatures. I shall return to this point below, and then complete the demonstration by comparing the results at which I have arrived with those predicted by the theory of M. Hirn.

Recently M. Winkelmann has published the results of experiments made with ethylene between 0° and 100° and between 1 and 2 atmospheres. He finds, corroboratively, that at 100° ethylene obeys the law of Mariotte more nearly than at 0° C. This is precisely my deduction for all the gases treated up to 100°, 250°, and 320°.

With the exception of these experiments and those described in the classical memoir of Andrews on the critical point, I believe that no other experimental data are available, barring the results which I adduced in the memoir summarizing my experiments.

As to laws treating of the manner in which the compressibility of a gas is modified at high pressures throughout different temperatures, they are quite unknown. It was, therefore, with the purpose of filling this important gap that I undertook the researches which make up the subject of the present memoir.

I have studied the gases nitrogen, hydrogen, marsh gas, ethylene, and carbon dioxide. As for air, and particularly oxygen, I fear that at the higher temperatures the action of the latter gas on mercury will be too rapid to admit of the necessary measurements, and this in greater degree as the experiments are more prolonged under the conditions of high temperatures stated.

At ordinary temperatures, on the other hand, the experiments can be conducted with relatively great despatch. A series of six determinations may be made with rigor in a quarter of an hour, and then all oxidation is inappreciable. Hence, in my first research I was able to study these two gases (by the method of comparison) without being annoyed by the difficulty in question. It is my object, moreover, to make a special study of it at some other time.

DESCRIPTION OF THE APPARATUS

The apparatus of which I here make use is the same which has already served me in the determination at ordinary temperatures of the compressibility of the gases air, oxygen, hydrogen, carbon dioxide, marsh gas, and ethylene, in comparison with nitrogen. I have briefly referred to it in my first memoir without giving a full description. However, as on that occasion I entered with considerable detail into the construction of divers parts of this class of instruments, I am able to confine myself to narrower limits here.

Figure 1 shows the apparatus drawn to a scale of one-twelfth actual size. On the right hand is the nitrogen manometer.* It has the identical dimensions and is mounted in quite the same manner as the one on the machine described in my first memoir on the subject. In this case it was used to measure the compressibility of nitrogen in the Verpilleux mine shaft. The jacket through which the current of water circulates and the enclosed thermometer are attached in the same way. The lid or flange which carries the whole is secured by four bolts to a hollow block of cast-iron, in the cavity of which the reservoir of the manometer is situated. On one side is a small wheel by which a cylindrical plunger, passing through a long box stuffed with leather, may be actuated for the purpose of regulating the applied pressure or of bringing the mercury meniscus in the stem of the manometer upon any division mark determined in advance. This pressure was calculated from the known compressibility of nitrogen.

On the left hand of the figure is another massive hollow block carrying a manometer, within which is the gas to be compared with nitrogen. This part is more bulky than the former, and as a whole similar to the apparatus described in my first memoir. Indeed, the experiments then treated might in any case be repeated with it. The difference lies merely in details of construction, and principally in the disposition of the stopcocks. Only the right half of this large block is to be used in the experiments with which we are here concerned. The left half of this part of the figure, carrying a lid with three bolts, may be overlooked. In the earlier manometric experiments the lower end of the hollow steel filament which ran to the top of the mine shaft was here affixed. [*See couplings attached.*]

* In the *Comptes Rendus*, April 12, 1880, I showed that under the influence of pressure applied in the interior only these manometers are not subject to an increase of volume serious enough to make special corrections necessary.

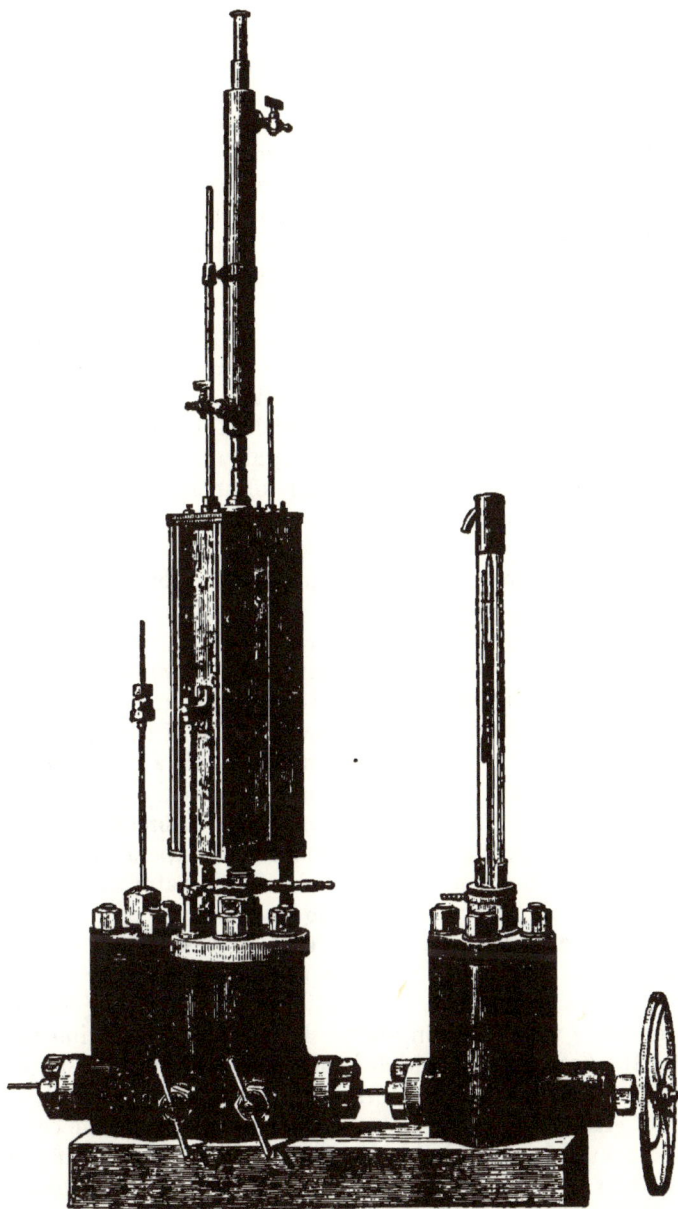

FIG. 1.—APPARATUS FOR THE COMPRESSION OF GASES AT DIFFERENT TEMPERATURES

B

If the observations are to be made at atmospheric temperatures, the manometer containing the second gas is jacketed in exactly the same way as the nitrogen manometer. If, however, the observations are to be made at higher temperatures, the glass tube with circulating water is replaced by a rectangular trough, or vapor bath, supported on three iron columns, screwed to the lid of the same block of iron which holds the manometer. For this reason the lid is circular and a little too large, jutting out slightly beyond the block to which it is bolted.

The trough was cast of a single piece of brass. On opposed faces it is provided with two long and narrow windows, closed with plate-glass held in place by brass plates suitably screwed to the sides of the trough. The top is closed with a square brass lid, also secured in place with bolts. To this is screwed a vertical condenser, the purpose of which is to liquefy all escaping vapor and to return it to the trough.

The condenser also carries the slide of an agitator or stirring device for equalizing the temperature of any liquid in the trough.

Temperatures may be read off on the thermometer inserted through and supported by the lid. As is seen, the trough is very solidly put together. It could, if necessary, resist great external pressures or be used for experiments *in vacuo*. The glass plates were not fastened with red-lead cement, as is usually done, but the hermetic seal was made by aid of thick plates of vulcanized rubber. I cannot recommend this adjustment too highly, as it is more convenient and more quickly put together than a red-lead joint and much less liable to break the plate-glass windows throughout the limits of temperature within which india-rubber can be employed. With water at 100° this method of sealing leaves nothing to be desired.

The stem of the manometer of crystal glass passes through the bottom of the trough. The joint is easily made by aid of a tubulure and a perforated cork sliding on the stem. It may also be made with a stuffing-box. In the actual experiments I prefer the first method, which is more simple and gives a good hermetic joint when the necessary precautions are taken, the stem of glass being made to pass through exactly at the centre of the tubulure which holds the cork. The trough, being adjustable by aid of double nuts on the columns, easily admits of any desirable change of position. One of the columns carries a movable ring-burner, through the centre of which the stem of the manometer passes. The crown of flame thus obtained suffices to heat the liquid in the trough to any required degree.

The two parts of the machine are connected by a short tube of steel. The apparatus on the left communicates with the mercury pump mentioned in my first memoir through a somewhat longer piece of the same steel tubing. As the pump has been already described, it needs no further consideration here.

The two couplings or stopcocks seen near the bottom of the apparatus on the left of the figure serve to put into communication different parts of the apparatus, either with themselves or with the pump, or to interrupt these communications. There is still a third coupling in the extreme left of the figure, but this is not essential in the actual experiments.

The whole apparatus is fixed to a heavy block of wood by aid of flanges of iron, which I have omitted in the figure. The wooden block in turn is secured to a truck with four rollers of cast-iron ; this allows an easy motion of the whole apparatus, whose weight is quite considerable. Four strong levelling screws were provided for adjusting the stems of the manometers vertically, giving to the whole mechanism additional stability.

METHOD OF EXPERIMENTATION

Having partially filled the apparatus with pure and dry mercury, the nitrogen manometer is first installed, and thereafter the second manometer provided with its trough. As the latter piece is very heavy, it would be difficult to adjust it manually without accident. I therefore had a double differential pulley attached to the ceiling of my laboratory, permitting me to accomplish this operation easily and without danger. Having completed the adjustment of the apparatus, mercury is injected by the pump until this liquid shows itself in the stems of the manometers. The stopcock on the left is now closed, while additional pressure is applied through the small hand-wheel, the stopcock on the right being left open.

To begin the measurements, the water in the trough is heated to the temperature selected, and the necessary constancy of temperature is maintained by suitably regulating the flow of gas and the distance of the burner from the bottom of the trough. This operation unfortunately is often very prolonged, particularly at the higher temperatures. During the whole time the stirring device must be kept in uniform action.

When constancy of temperature has been reached, a heavy table of wood is placed facing the apparatus, and two telescopic sights, fixed to the table, are directed to the meniscuses in the two manometers respectively. Thereupon an assistant, by means of the hand-wheel, brings the mercury successively on ' the division lines of the nitrogen manometer corresponding to the pressures calculated in advance. The observer at the telescopes registers the indications of the two manometers simultaneously with those of the thermometer of the trough and of the jacket of water circulating around the nitrogen manometer.

As to the latter, observation is made very simply by placing a white screen behind the tube, and always bringing the base of the meniscus upon a division mark of the scale. To read the other manometer, a luminous background, preferably the ground-glass globe of a gas-lamp, is placed at some distance behind the posterior window. The reading of the summit of the meniscus is then observed, well marked by its demarcation from the bright field. In estimating volumes with the aid of a prepared calibration-table, allowance is made for this difference of positions of the meniscuses in the readings.

METHOD OF CALCULATION

The experiments which I have made with the different gases are all comprised between 30 and 420 atmospheres, and between atmospheric temperatures and 100° C.

Each gas was studied in two series of observations. With the first manometer the experiments were carried as far as 100 or 130 atmospheres ; with the second I proceeded from 100 or 130 atmospheres as far as 420 atmospheres. In this way too great a reduction of the volume of gas is avoided.

The measurement of pressure given by the nitrogen manometer was also carried out with two pieces of apparatus for like reasons.

This procedure renders the measurement of volume more

exact, but it has the drawback of presenting difficulties in recording the indications of the two manometers and in reducing them to a single series. I made this reduction by two different methods, corroborating them as I shall indicate below.

The curves representing the results have been constructed as follows : the abscissas are laid off proportional to the pressures, .millimetres of length corresponding to meters of mercury; the lengths of the ordinates are proportional to the corresponding values of the products pv of pressure and volume.

All the isotherms for the same gas at the different temperatures refer to the same gaseous mass ; in other words, the manometers, after once being charged with a given gas, were used in the experiments throughout all temperatures without taking the apparatus apart. To record these families of curves as a whole, I made use of the following two procedures : Consider any particular curve. Supposing that the first manometer has furnished the products pv as far as 130 atmospheres, for example, and the second as far as 120 atmospheres, deduce from the curve constructed up to 130 atmospheres the value of pv at 120 atmospheres. All the numbers furnished by the second manometer are now multiplied by the ratio of the value of pv at 120 atmospheres taken from the first curve to that of the same product furnished by the second manometer at the same pressure. This factor, once calculated, is sufficient to co-ordinate the relative curves at all temperatures, on condition, let it be well understood, that the series which are thus co-ordinated have been made with each manometer at exactly the same temperature. These circumstances were always realized to about 1° C., and the coefficient calculated for one of the curves has always made the others agree. To insure greater accuracy I computed the coefficient for all the temperatures and selected the mean value. Thereafter I proceeded as follows : Suppose the gas introduced successively into the two manometers rigorously at the same temperatures and at the same pressure. It would then obviously be sufficient to multiply all the products pv furnished by the second manometer by the ratio of the total volume of the first manometer to that of the second. In reality, however, it was necessary to make a small correction, due to the difference of temperatures and of pressures at which the manometers were charged; but this correction was in all cases very small. The coefficients determined in this way agreed with sufficient accuracy with those determined by the first procedure. They were usually found to be equal to about one per cent., and the mean was taken. Frequently the agreement was even within these limits.

Finally the curves drawn to the scale which I have given above were reduced to one-third of their original dimensions, which far exceeded the usual size for publication.

The observations furnished at the nitrogen manometer were reduced in the following way : A table of products pv was first prepared by the aid of my earlier researches, in which the initial volume is considered identical with that of the manometer, and the initial pressure that under which it was charged. Therewith a large curve was constructed by laying off volumes along the abscissas and the values of pv along the ordinates. From this I deduced the values of pv for the volumes corresponding to the scale marks on the nitrogen manometer, up to which the mercury meniscus was forced. Dividing these values by the corresponding volumes deduced from the tables of calibration, the pressure-values sought are obtained. Thus it is supposed that the reading is made at the same temperature as that at which the manometer was charged at normal pressure. Hence the correction relative to the difference of temperature is always very small. I have also computed the pressures by another method, which gave me essentially the same results. To corroborate the whole I placed the two nitrogen manometers simultaneously upon the pressure apparatus, and I thus found that throughout the coincident part of their registry—$i. e.$, between about 100 and 130 atmospheres—they gave the same results to about half an atmosphere.

It has just been shown that in making the calculations needed for joining the curves it is necessary to know at what temperature and at what pressure the manometers were charged with gas. The experiments were so conducted that the small cylinder which ends off the reservoirs of the manometers below was submerged in mercury at the moment at which its pressure and its temperature were identical with that of the atmosphere, although communication with the gas supply had not yet been cut off. While the small cylinder was thus plunged in the mercury bath, it was provided with a small iron thimble, kept in place by suspension from two springs. In this way the manometer could be removed or placed upon the apparatus, the efflux point remaining plunged in the mercury within the thimble.

As a general thing, three series of operations were made with each gas and with each manometer. Having thus established the accordance of a large number of series, I selected the one which seemed to have been made under the most satisfactory conditions.

I took every feasible precaution to have the gases as pure, as free from air, and as dry as possible. For the case of carbon dioxide, in particular, the gas was tested even in the manometer at the end of the experiments. The results given below refer to a gas which left a residue of only 5 parts in 1000 in the lower-pressure manometer and 2 parts in 1000 in the other. All the gases, without exception, were first dried in concentrated sulphuric acid, thereafter in a desiccator filled with broken glass mixed with anhydrous phosphoric acid.

NUMERICAL RESULTS

The numerical results contained in the following tables were obtained from the curves which have already been discussed. For ethylene and carbon dioxide, in which the variations of pressure are exceedingly abrupt, I have given the values of pv in intervals from 10 to 10 meters of mercury; for the other gases, in intervals of 20 meters only. The first horizontal row shows the temperatures at which the observations were made. The corresponding pressures are contained in the first vertical column.

In case of marsh gas (methane) data are given only up to 300 atmospheres. This gas was actually studied in certain series as far as 420 atmospheres, like the others; but an error appeared in the values of the coefficient for co-ordinating these curves which increased to more than three per cent., and which I have not been able to explain to my own satisfaction. I therefore prefer to give the data of the first series only, which showed sufficient accordance throughout.

VALUES OF pv FOR NITROGEN

Meters, Hg.	17.7°	30.1°	50.4°	75.5°	100.1°
30	2745	2875	3080	3330	3575
40	2740	2865	3085	3340	3580
60 ·	2740	2875	3100	3360	3610
80	2760	2895	3125	3400	3650
100	2790	2930	3170	3445	3695
120	2835	2985	3220	3495	3755
140	2890	3040	3275	3550	3820
160	2950	3095	3335	3615	3880
180	3015	3150	3390	3675	3950
200	3075	3220	3465	3750	4020
220	3140	3285	3530	3820	4090
240	3215	3360	3610	3895	4165
260	3290	3440	3685	3975	4240
280	3370	3520	3760	4050	4320
300	3450	3600	3840	4130	4400
320	3525	3675	3915	4210	4475

VALUES OF pv FOR HYDROGEN

Meters, Hg.	17.7°	40.4°	60.4°	81.1°	100.1°
30	2830	3045	3235	3430	3610
40	2850	3065	3240	3445	3625
60	2885	3110	3295	3500	3680
80	2935	3155	3340	3550	3725
100	2985	3200	3400	3620	3780
120	3040	3255	3455	3665	3830
140	3080	3300	3500	3710	3880
160	3135	3360	3560	3775	3945
180	3185	3420	3620	3830	4010
200	3240	3465	3685	3870	4055
220	3290	3520	3725	3930	4110
240	3340	3570	3775	3980	4160
260	3400	3625	3830	4040	4220
280	3450	3675	3880	4090	4275
300	3500	3730	3935	4140	4325
320	3550	3780	3990	4200	4385

THE LAWS OF GASES

MARSH GAS (METHANE)

Meters, Hg.	14.7°	29.5°	40.6°	60.1°	79.8°	100.1°
30	2580	2745	2880	3100	——	——
40	2515	2685	2830	3060	3290	3505
60	2400	2590	2735	2995	3230	3460
80	2315	2515	2675	2950	3195	3440
100	2275	2480	2640	2935	3180	3435
120	2245	2465	2635	2925	3180	3440
140	2260	2480	2655	2940	3190	3460
160	2300	2510	2685	2975	3220	3490
180	2360	2560	2730	3015	3260	3525
200	2425	2615	2780	3065	3305	3575
220	2510	2690	2840	3125	3360	3625
230	2560	2730	2880	3150	3385	3650

VALUES OF pv FOR ETHYLENE

M. Hg.	16.3°	20.3°	30.1°	40.0°	50.0°	60.0°	70.0°	79.9°	89.9°	100.0°
25	2140	2215	2360	——	——	——	——	——	——	——
30	1950	2055	2220	2410	2580	2715	2865	2970	3090	3225
40	1350	1700	1900	2145	2335	2510	2675	2825	2960	3110
50	850	1075	1540	1860	2100	2315	2490	2670	2825	2980
60	810	900	1190	1535	1875	2100	2310	2500	2680	2860
70	880	945	1110	1340	1675	1920	2150	2365	2560	2740
80	975	1030	1130	1285	1535	1780	2015	2240	2450	2640
90	1065	1115	1195	1325	1510	1710	1930	2160	2375	2565
100	1150	1200	1275	1380	1535	1690	1895	2105	2335	2515
110	1240	1280	1360	1460	1590	1725	1915	2705	2310	2490
120	1325	1370	1440	1540	1660	1780	1950	2115	2305	2470
130	1415	1455	1525	1620	1725	1840	2000	2150	2320	2480
140	1505	1540	1610	1700	1800	1910	2060	2190	2350	2505
150	1590	1625	1690	1785	1880	1990	2125	2250	2390	2540
160	1680	1715	1780	1865	1960	2070	2195	2310	2445	2585
170	1770	1800	1860	1950	2045	2145	2265	2375	2505	2640
180	1855	1890	1945	2035	2130	2225	2340	2450	2565	2700
190	1940	1975	2030	2120	2210	2310	2415	2525	2635	2760
200	2030	2065	2115	2200	2290	2390	2490	2600	2715	2835
210	2110	2145	2200	2285	2375	2470	2565	2680	2790	2910
220	2195	2225	2280	2370	2460	2550	2650	2760	2865	2975
230	2280	2315	2370	2460	2540	2635	2730	2835	2940	3050
240	2360	2395	2450	2540	2625	2720	2810	2910	3015	3125
250	2445	2480	2540	2625	2710	2800	2890	2990	3090	3200
260	2530	2560	2625	2710	2790	2880	2980	3075	3175	3275
270	2610	2640	2710	2790	2875	2965	3060	3150	3240	3345
280	2695	2725	2790	2875	2960	3045	3140	3225	3320	3420
290	2780	2810	2875	2960	3040	3125	3220	3310	3400	3490
300	2860	2890	2960	3040	3125	3215	3300	3380	3470	3560
310	2945	2975	3040	3125	3210	3290	3385	3465	3550	3635
320	3035	3065	3125	3200	3285	3375	3470	3545	3625	3710

VALUES OF pv FOR CARBON DIOXIDE

M. Hg.	18.2°	35.1°	40.2°	50.0°	60.0°	70.0°	80.0°	90.2°	100.0°
30	Liquid	2360	2460	2590	2730	2870	2995	3120	3225
40	"	2065	2195	2370	2535	2700	2840	2985	3105
50	"	1725	1900	2145	2330	2525	2685	2845	2980
60	"	1170	1500	1860	2115	2340	2530	2705	2860
70	"	725	950	1530	1890	2155	2380	2570	2750
80	625	750	825	1200	1650	1975	2225	2440	2635
90	685	810	865	1080	1430	1775	2075	2315	2530
100	760	870	920	1065	1315	1630	1940	2200	2425
110	825	930	980	1090	1275	1550	1845	2105	2325
120	890	995	1045	1140	1285	1510	1775	2030	2260
130	955	1060	1115	1190	1315	1505	1735	1980	2190
140	1020	1120	1175	1250	1360	1525	1715	1950	2160
150	1080	1180	1235	1310	1415	1560	1725	1945	2135
160	1145	1250	1300	1370	1465	1600	1745	1960	2130
170	1210	1310	1360	1430	1520	1645	1780	1975	2135
180	1275	1375	1410	1485	1580	1700	1825	2000	2150
190	1340	1440	1480	1550	1645	1760	1875	2035	2180
200	1405	1500	1550	1615	1705	1810	1930	2075	2215
210	1470	1565	1610	1675	1765	1870	1980	2120	2250
220	1530	1625	1670	1740	1825	1925	2040	2160	2290
230	1590	1690	1730	1800	1890	1990	2090	2210	2340
240	1650	1750	1790	1865	1950	2045	2150	2260	2390
250	1710	1815	1855	1925	2010	2100	2205	2320	2435
260	1770	1870	1920	1985	2070	2165	2265	2375	2490
270	1830	1935	1975	2050	2130	2220	2320	2435	2540
280	1890	2000	2040	2110	2190	2285	2380	2490	2600
290	1950	2060	2100	2170	2260	2340	2440	2550	2655
300	2010	2120	2160	2235	2320	2405	2500	2605	2715
310	2070	2180	2220	2300	2375	2460	2560	2660	2765
320	2135	2240	2280	2360	2440	2525	2620	2725	2830

The curves given in figures 2, 3, 4, 5, 6, represent these results comprehensively. In those relative to nitrogen, to hydrogen, and to marsh gas a part of the ordinate lengths has been suppressed. It is therefore necessary to reproduce this inferentially with the aid of the numbers given at the origin.

EXAMINATION AND DISCUSSION OF THE RESULTS

An inspection of the curves shows at once that the families may be referred to two extreme and to certain intermediate types. For hydrogen all the lines are appreciably parallel and straight at all the temperatures at which experiments were made. This invariability in the form of the curves seems to indicate that this gas has reached a limiting state characterized by their direction. At all the temperatures which I investigated the values of pv increase in their variation with pressure.

Carbon dioxide and ethylene form the contrasting type. The products pv at first decrease very rapidly, reach a minimum,

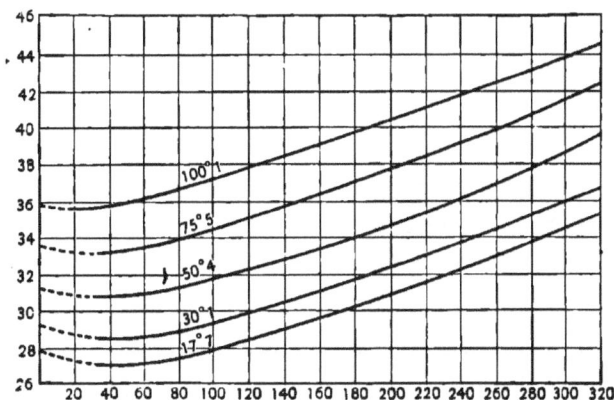

Fig. 2.—Isotherms (pv) for Nitrogen

and thereafter increase indefinitely. These variations of pv, very rapid at temperatures near the critical point, show a marked diminution when temperature rises. The point of the curve at which the ordinate is a minimum moves regularly away from the origin, and the locus traced comes out very clearly on inspection of the curves. The minimum seems to move away from the origin less rapidly after passing a certain temperature ; after which it apparently retrogrades. At least, this takes place in marsh gas and nitrogen. Now these gases which constitute the intermediate type are at like temperatures much more distant from their critical state than ethylene or carbon dioxide. For the case of nitrogen and marsh gas the displacement of the minimum ordinate appears much less sharply defined than for the other gases, for the reason that when curvature diminishes, the position of this ordinate is much more difficult to determine sharply. I subjoin a table for carbon dioxide and ethylene,

27

showing at what pressure in meters of mercury the ordinate is a minimum at different temperatures :

CARBON DIOXIDE		ETHYLENE	
—	—	16.3°	55^M
—	—	20.3	60
35.1°	70^M	30.1	70
40.2	80	40.0	80
50.0	98	'50.0	88
60.0	115	60.0	95
\ 70.0	130	70.0	100
80.0	140	79.9	105
90.2	150	89.9	115
100.0	160	100.0	120

The curves for carbon dioxide and ethylene may advantageously be considered by themselves for the time being ; because of the larger variations of compressibility involved, they

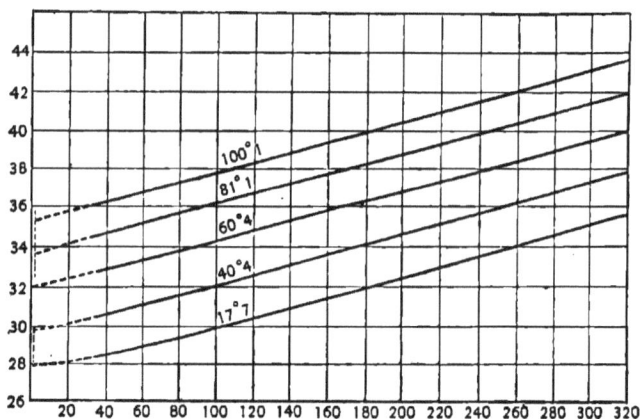

FIG. 3.—ISOTHERMS (pv) FOR HYDROGEN

are more suitable than the others for the demonstration of the general laws of these variations. Let the curves for carbon dioxide be taken first. · It is obvious at the outset that at temperatures in the neighborhood of the critical point the initial branches of the curves, or those which precede the minimum ordinate, are concave towards the axis of pressure. The concavity is well marked, and appears to be prolonged quite into the region of small pressures. This is indicated by the dotted lines which represent the phenomenon approximately, at pressures lower than those at which the experiments began.

To construct these parts of the curve I simply determined

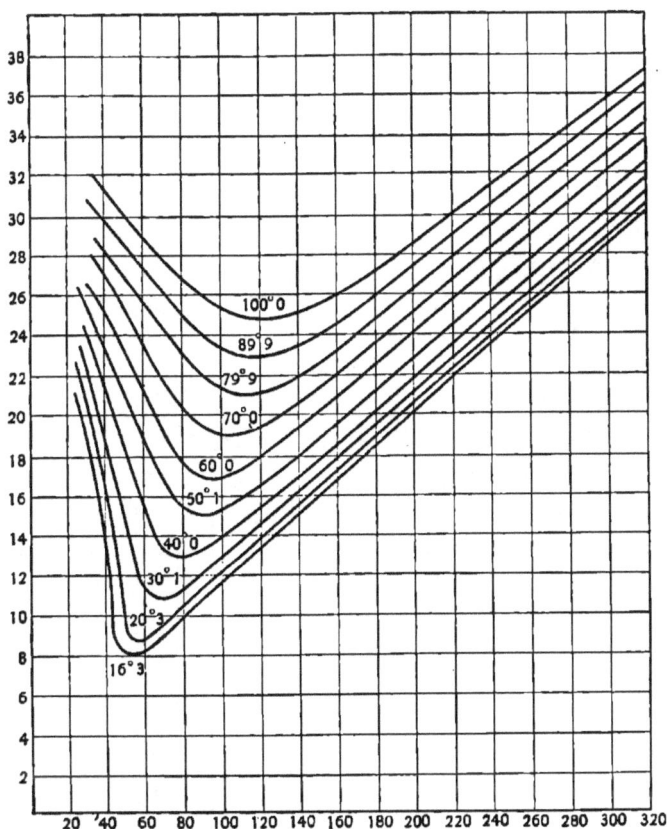

FIG. 4.—ISOTHERMS (pv) FOR ETHYLENE

the point of departure corresponding to normal pressure, which was found without difficulty, since I knew the volume of carbon dioxide for the particular pressure and temperature at which the manometer was charged with gas. From this I deduced the volume occupied by the gas, and consequently the value of *pv*, at the same pressure for all the temperatures, in virtue of the values of the coefficient of expansion of carbon dioxide, given a long time ago in my own papers, for temperatures between 0° and 100° and under normal pressure. Thereupon I joined the points of departure so obtained with the first points of the continuous curves, allowing myself to be guided by their

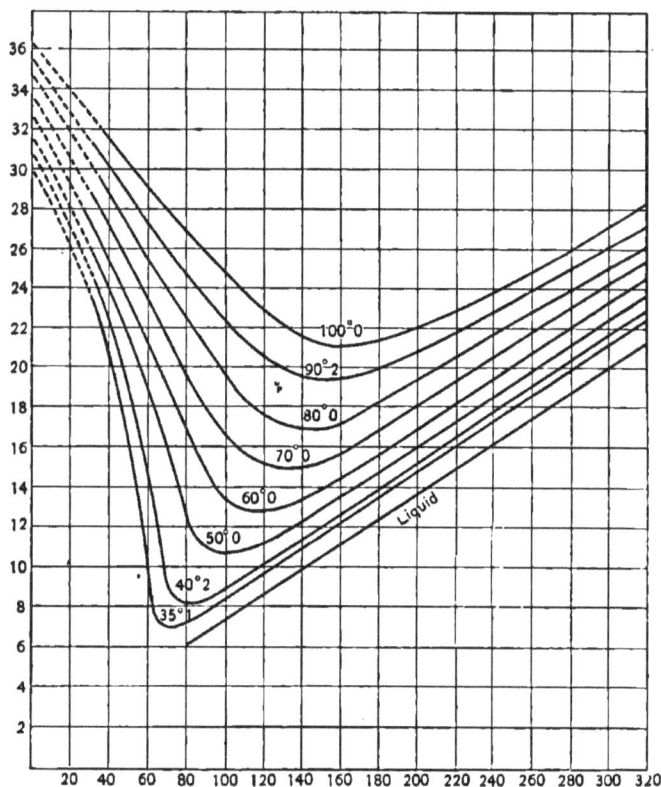

FIG. 5.—ISOTHERMS (*pv*) FOR CARBON DIOXIDE

general trend. Similar dotted lines were drawn for nitrogen and hydrogen, but not for ethylene and methane, since I lacked data as to their dilatation.

To return from this digression to the general form of the curves, I may point out that the concavity observed in the initial branches which is presented in the case of ethylene, occurs also, I might almost say *a fortiori*, in the results which Regnault investigated as far as 28 atmospheres. In fact, these data were obtained at temperatures lower than those from which I started—*i. e.*, lower than 35° for carbon dioxide.

The concavity specified vanishes pretty rapidly when temperature rises. At 50° C. it is no longer apparent, and the curve only presents the convexity which produces the minimum of *pv*.

For gases other than ethylene and carbon dioxide, the concavity has entirely disappeared even at ordinary temperatures. This form of curve for the lower pressures, for the moment at

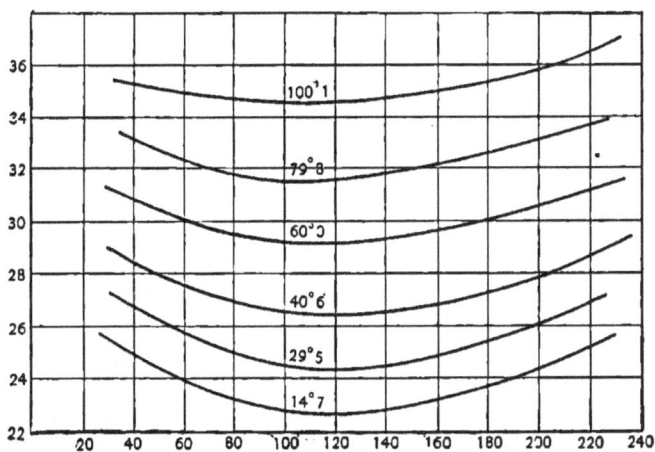

FIG. 6.—ISOTHERMS (*pv*) FOR MARSH GAS

least, does not command much attention; but it is quite different in those parts of the curves which follow the minimum ordinate, and throughout which *pv* increases indefinitely. For temperatures in the neighborhood of the critical point, the curve turns rapidly after passing the ordinate in question, and changes almost at once into a line which is very nearly straight. Some points seem to indicate traces of concavity, so vaguely, however, as to be referable to errors of observation.

As temperature rises the convexity of the curves diminishes very rapidly, and the general aspect of the families of curves shows an unmistakable tendency to slope upward and to change to straight lines throughout their extent. The occurrence of minimum ordinates thus gradually ceases.

True, this tendency is indicated clearly only in the isotherms of carbon dioxide and of ethylene. If, however, one calls to mind that hydrogen is at ordinary temperatures in a thermal state, which the two gases specified reach at very much higher temperatures; that for hydrogen the transformation in question is actually complete at ordinary temperatures, continuing in the same sense up to 100° without showing the slightest trace of any superimposed deformation; if, finally, one adds the evidence obtained from the curves for methane and nitrogen which make up the intermediate type, then the general features of these gaseous phenomena become strikingly apparent. Indeed, I would insist on this point of view (the consequences of which will be shown below), that not only do the curves straighten out for all gases in such a way as to reproduce the case of hydrogen at sufficiently high temperatures—*i.e.*, in a way to present values of *pv* continually increasing with pressure—but that they really tend to become *straight lines* throughout their whole extent.

For hydrogen the occurrence of straight isotherms is as completely verified as the accuracy of the experiments permits up to 100 atmospheres about and from ordinary temperatures onward. At smaller pressures, a trace of concavity seems still to cling to the curves. This is much less discernible at 100°, if it be not altogether wiped out.

The concavity of the curves for nitrogen is clearly marked at ordinary temperatures, but it is much less accentuated at 100°, where the rectilinear portions are already apparent.

Another fact not less important (as will appear below) is this, that the curves in their rectilinear parts are nearly parallel; parallel in such a way, moreover, that to obtain the general direction of these lines—a characteristic for each gas—it is sufficient to construct one of them near the critical point. Under critical conditions the lines become straight almost immediately after leaving the minimum ordinate. Nevertheless, it would be hazardous to affirm that these lines are absolutely parallel; they seem rather to merge gradually into straight lines, being asymptotic to a direction which differs very little in all the cases for the portion already sensibly straight in the curves constructed.

The endeavor must now be made to unravel general laws relative to the variation of compressibility with temperature.

We have seen that for each gas there exists a temperature beyond which pv increases continually with pressure. It is, however, inaccurate to state that with increasing temperature the gas continues to diverge more and more seriously from the law of Mariotte, in the sense of being less compressible. The contrary is the fact. Let the values of the ratio $\frac{pv}{p'v'}$ be examined between the limits p and p' of pressure and at different temperatures. It follows from the parallelism of the lines that in passing from a given temperature to another higher in the scale the values of pv and $p'v'$ increase by the same quantity. Hence the value of the ratio increases, since it is smaller than one.

All this is clearly shown in the following table containing values of $\frac{pv}{p'v'}$ between 100 meters and 320 meters of pressure and at different temperatures:

HYDROGEN

° C.	Ratio $\frac{pv}{p'v'}$.
17.7	0.830
40.4	.838
60.4	.845
80.1	.853
100.1	.856

Thus for increasing temperatures the gas expands more in accordance with Mariotte's law, for the simple reason that the constant difference of the products pv and $p'v'$ is added to the greater and greater values of the products. The reason is not that pv tends to become constant, a result which occurs for no gas whatever.

It is quite certain that, on sufficiently cooling hydrogen, this gas would eventually contract more than the law of Mariotte indicates. On increasing temperature the discrepancy would at first vanish and thereafter change sign. But the divergence, instead of continuing to increase negatively, would reach a maximum value, after which it would begin to approach unity, as is actually the case in the experiments.

Let the family of curves relative to carbon dioxide now be examined : They begin at 35° C.—*i.e.*, only a few degrees above the critical point. Following the values $\frac{pv}{p'v'}$ from degree to degree between the pressure limits 100 and 120 meters of mercury, for example, we find the ratio smaller than unity at 35°, 40°, and 50°. Above 60°, on the contrary, the ratio exceeds unity ; after this it evidently again becomes smaller as the temperature continually increases, and finally remains indefinitely at a point below this value. Clearly, therefore, the period within which the gas is more compressible than the law of Mariotte asserts may be preceded by a period where it is less so at a lower temperature. The curves also show that a pressure exists beyond which $\frac{pv}{p'v'}$ is always smaller than unity, whatever be the temperature. If one imagines the locus of the points of minimum ordinates actually constructed, and if the curve has itself a maximum abscissa (I have shown that this is very probable), it is beyond the pressure corresponding to this abscissa that the fact in question becomes manifest.

Approaching the region of lower pressures one observes similarly that a value exists beyond which the primary period corresponding to $\frac{pv}{p'v'} < 1$ no longer occurs. The value in question is the pressure at which the gas liquefies at a temperature very near the critical point. It is the critical pressure.

All these results are rather complicated, although an inspection of the curves immediately shows them clearly. They may, however, be enunciated in a simpler form. Suppose the family of curves divided into two regions by the curve which is the locus of minimum ordinates. The left-hand area will then comprehend those parts of the curves for which $\frac{pv}{p'v'} > 1$; in the right-hand area, contrariwise, we shall everywhere have $\frac{pv}{p'v'} < 1$. In the first region, $\frac{pv}{p'v'}$ decreases when temperature increases ; in the second the opposite is the case.

The following table has been computed from the data for carbon dioxide to substantiate these results. It may be observed in passing that in the second region the effect of temperature diminishes whenever pressure increases.

34

CARBON DIOXIDE

TEMP.	VALUES OF $\dfrac{pv}{p'v'}$				
	30M—70M	70M—100M	100M—140M	140M—200M	200M—320M
18.2° (Liq.)	—	—	.745	.726	.658
35.1 "	3.255	.833	.777	.747	.670
40.2 "	2.893	.924	.783	.758	.688
50.0 "	1.693	1.452	.844	.774	.685
60.0 "	1.444	1.437	.967	.797	.699
70.0 "	1.329	1.322	1.069	.842	.712
80.0 "	1.258	1.227	1.131	.888	.736
90.2 "	1.214	1.168	1.128	.940	.761
100.0 "	1.172	1.134	1.122	.975	.782

The limits of pressure between which $\dfrac{pv}{p'v'}$ has been calculated are inscribed at the head of the respective vertical columns. We may therefore summarize the role of temperature in its effect upon the compressibility of gases by the following laws :

1. For pressures lower than the critical pressure and with continually increasing temperature, the divergence from Mariotte's law, positive at first at sufficiently low temperatures, passes through zero and eventually becomes negative. Beyond a certain negative value, however, the discrepancy diminishes indefinitely without changing sign.

2. For pressures between the critical value and a superior limit peculiar to each gas, the period during which the discrepancy is positive is preceded at still lower temperatures by a period for which it is negative, in such a way that the discrepancy changes sign twice.

3. Beyond the superior limit indicated in the preceding law the discrepancy is negative at all temperatures. It diminishes in general as temperature increases, always excepting those pressures which are too near the limit specified. In these places the variation is more complicated.

4. Beyond a sufficiently high temperature the law of compressibility of a gas is represented by the equation $P\,(V-a)=$ const., wherein a is the smallest volume to which the gas can be reduced; in other words, a is the absolute volume of the constituent matter.

35

The last law virtually states (as will appear below) that beyond a certain sufficiently high temperature all the curves become straight lines.

It goes without saying that the departure from the law of Mariotte here in question refers to pressures arbitrarily chosen within the limits of pressure indicated by the laws.

DILATATION OF GASES AT HIGH PRESSURES

The data which precede are evidently available for the computation of the coefficient of expansion of a gas, even though the experiments were not specially directed towards this end. True, results so obtained cannot have a degree of precision comparable with those investigated by the ordinary methods at low pressures; but the accuracy will nevertheless be sufficient to point out the general features.

Mere inspection of the families of curves enables us to form a conception of the remarkable variation to which the dilatation of a gas is subject in the neighborhood of the locus of minimum ordinates; above all, at temperatures near the critical point.

If we reflect that for a given pressure the length of the ordinate is at each temperature proportional to the volume of the gas, it follows that the dilatation for each pressure is consequently given by the curves. To arrive at the general facts, however, it is necessary to compare the coefficients of mean expansion at different pressures between the sufficiently narrow limits of temperature. It is with this end in view that I have computed the table which is about to follow.

The coefficients of expansion inserted are the values of $\dfrac{v'-v}{v\,(t-t')}$ between the limits of pressure and of temperature indicated. The volumes were deduced from the curves by dividing the ordinates pv by the corresponding pressure. It sufficed for this purpose to take the difference of ordinates corresponding to t and t' degrees at each pressure, to divide this difference by the inferior ordinate giving $\dfrac{pv'-pv}{pv}$ or $\dfrac{v'-v}{v}$, and finally to divide this result by the difference of temperature.

The data thus obtained for carbon dioxide and ethylene follow:

CARBON DIOXIDE

Pressure	18°—35°	40°—60°	60°—80°	80°—100°
40 Meters	Liquid	.0074	.0058	.0046
60 "	—	.0196	.0096	.0052
80 "	.0113	.0500	.0176	.0089
100 "	.0072	.0217	.0238	.0135
120 "	.0062	.0114	.0151	.0123
140 "	—	.0085	.0128	.0127
160 "	.0043	.0066	.0095	.0108
180 "	—	.0056	.0079	.0087
200 "	.0039	.0052	.0071	.0072
220 "	—	.0048	.0057	.0063
240 "	.0033	.0045	.0051	.0056
260 "	—	.0040	.0045	.0048
280 "	.0029	.0039	.0042	.0046
300 "	—	.0038	.0039	.0044
320 "	.0025	.0037	.0038	.0040

ETHYLENE

Pressure	20°—40°	40°—60°	60°—80°	80°—100°
30 Meters	.0084	.0064	.0046	.0040
60 "	.0366	.0178	.0097	.0067
80 "	.0121	.0195	.0132	.0088
100 "	.0079	.0108	.0121	.0100
120 "	.0062	.0075	.0095	.0082
140 "	.0048	.0062	.0076	.0068
160 "	.0041	.0057	.0061	.0058
200 "	.0034	.0043	.0044	.0044
240 "	.0030	.0035	.0036	.0034
280 "	.0027	.0031	.0030	.0029
320 "	.0025	.0027	.0024	.0024

If by running along a horizontal column one endeavors to find how the coefficient of expansion varies with temperature at constant pressure, one observes a rather complex result for the values given in the middle of the table, which correspond to the neighborhood of the locus of minimum ordinates; but if one considers only the extreme regions, regions which correspond to low pressures or to pressures relatively high, it becomes easily manifest that the coefficient diminishes regularly with temperature. Particularly on arriving near the limits of pressure will it be observed that the expansion is sensibly proportional to the interval of temperature. This is the case with hydrogen at all pressures.

Again, if in a survey of the vertical columns of the tables we endeavor to find the variation of the coefficient of expansion with pressure at a given constant temperature, we encounter a clear-cut law at once. The coefficient is at first seen to increase with pressure up to a maximum value, and thereafter to decrease regularly. This maximum corresponds very nearly to the pressure for which the ordinate is a minimum. If in place of taking the mean coefficient between limits as far apart as 20° the temperature interval $t'-t$ be more and more diminished, the limiting value $\dfrac{dv}{dt}\dfrac{I}{v}$ will coincide as to the pressure position of its maximum with the same pressure which corresponds to the minimum ordinate.

In proportion as temperature increases, this maximum is less and less marked until it eventually vanishes, as in the case of hydrogen. The following table, drawn up for this gas, shows at the same time that the mean coefficient decreases uniformly when pressure increases.

HYDROGEN

Pressure	17°—60°	60°—100°
40 Meters	.0033	.0029
100 "	.0033	.0028
180 "	.0031	.0027
260 "	.0030	.0025
320 "	.0028	.0024

I may, therefore, summarize the laws relative to the expansion of gases in the following way:

1. *The coefficient of expansion of gases (referred to the unit of volume) increases with pressure up to a maximum value, beyond which it decreases indefinitely.*

2. *The pressure corresponding to this maximum coincides in position with the pressure for which the product pv is a minimum. Consequently, at this exceptional point the gas accidentally obeys the law of Mariotte.*

3. For continually increasing temperatures the maximum in question becomes more and more indistinct and finally vanishes.

COVOLUME—ATOMIC VOLUME

We have seen that for hydrogen the curves obtained are nearly straight lines, and that the same is the case for carbon dioxide and for ethylene throughout a considerable part of the region beyond the minimum ordinate. We have seen furthermore that for increasing temperatures the curves tend more and more to become straight lines throughout their whole extent, thus again resembling the phenomenon observed with hydrogen at temperatures above that of the atmosphere. In this case the curves become

(1) $$pv = a\,p + b.$$

The initial ordinate b being the value of pv at the limit—*i.e.*, for a pressure infinitely small, if the laws which we have adduced are true under these conditions. This indeed is a question which has not yet been sufficiently studied and which I shall shortly take up again. We shall therefore regard our inferences limited as to pressures to an interval within which the results present a sufficient degree of certitude. As to hydrogen, I have already stated that for pressures less than 100 atmospheres the line still shows a slight curvature even at 100° C. But it is reasonable to admit that this curvature will quite vanish at temperatures taken high enough, and that the line will become straight—at least, above the pressures in the neighborhood of normal pressure.

The equation (1) may be put in the form

(2) $$p\,(V - a) = b,$$

b and a being constants. The result arrived at is therefore this, that at a given temperature the product of the pressure and the volume diminished by a constant quantity does not vary. If furthermore the relation is written

$$(V - a) = \frac{b}{p},$$

it appears that when $p = \infty$, $v = a$. That is, a is the volume which the gas eventually takes when pressure increases indefinitely. This may be rationally interpreted by considering a as the absolute volume of the matter within the gas, supposing that the molecules will ultimately touch each other, and not the molecules only but the atoms which make up those molecules.

Dupré and M. Hirn have reached a similar conclusion within certain limits by different methods, and in a way quite unlike that which I have just explained ; but their inferences would lead to interpretations which are at variance with my researches, taken as a whole.

In fact Dupré, from fundamental formulæ in the mechanical theory of heat, which he treats in his work (Dupré, *Théorie mécanique de la chaleur*), deduces the following law which he calls the law of *covolumes*, as an approximation of a higher order of accuracy than the law of Mariotte. I will quote it verbatim:

"*At constant temperature the pressures of a mass of gas vary inversely, as the volumes diminished throughout by a small constant quantity $c_o u_o$. This is to be called the covolume when the volume u_o under normal conditions is the unit of volume.*"

For nitrogen, carbon dioxide, and air the covolume of Dupré is positive ; for hydrogen it is negative. Seeking a verification of this law by aid of the numerical data in the classical research of Regnault, Dupré found that this was feasible for hydrogen, as may well be anticipated after what has just been said ; but for nitrogen, and above all for carbon dioxide, the verification leaves much to be desired. It could not be otherwise, since the law of the covolume presupposes that the curve representing the results in the above co-ordinates is straight. This condition is not realized for the case of nitrogen, nor for carbon dioxide, unless it be for temperatures and under pressures for which the covolume becomes precisely contrary in sign to that deduced by Dupré. The law of the covolume with the interpretation given to it by this physicist cannot therefore be admitted.

M. Hirn has published an elaborate research * on the same subject. Endeavoring to interpret the variations from the law of Mariotte, he points out that even if the molecules of a gas exert no reciprocal action on each other the gas cannot rigorously obey this law, since the variable part of the volume is not the total volume of the gas, but rather the latter diminished by the volume of the atoms. This is equivalent to admitting the quantity a defined above. M. Hirn contends that it is merely the variable part of the volume which ought to enter into the expression of Mariotte's law, a conclusion far from evident.

For the case in which one may not neglect the interactions

---

* Hirn : *Théorie mécanique de la chaleur*, t. ii.

THE LAWS OF GASES

of the molecules, M. Hirn introduces an internal pressure to be added algebraically to the external pressure, and the general expression of the law for constant temperature becomes

$$(P+p)\ (V-a)=(P'+p')\ (V'-a),$$

p and p' being the internal pressures corresponding to the volumes V and V'. For hydrogen p and p' would be approximately zero, whence

$$P\ (V-a)=\text{const.},$$

an expression which M. Hirn verifies by aid of the data of Regnault.

For nitrogen this correction of volume becomes insufficient, and does not even retain the same sign. It is therefore necessary, in order to explain the variation of this gas, to admit an internal pressure of marked value. Now if this is the case with nitrogen for pressures less than 20 atmospheres, one cannot assuredly deny that it must also be the case for hydrogen when this gas is reduced to the three-hundredth part of its volume. But for pressures as large as 430 atmospheres, and very certainly even for higher pressures, the law for the compressibility of hydrogen is given by the equation

$$p\ (V-a)=\text{const.}$$

as rigorously as for smaller pressures. The internal pressure is therefore zero. Furthermore, if we examine the data relative to carbon dioxide and ethylene, we again observe that for temperatures near the critical point a large part of the curve becomes straight. This takes place for carbon dioxide at 35° from 100 to 430 atmospheres; and the same phenomenon is observed at 18° C. between the same limits of pressure, even though the body is now a liquid—*i.e.*, has been subjected to what is properly termed liquefaction. Under these conditions the compressibility of the gas is thus represented by the relation

$$p(V-a)=\text{const.},$$

and a has the same value as at 100° C., and higher temperatures where the same formula represents the phenomenon from the lowest pressures upward. Hence one may argue that under circumstances in which the internal pressure ought to attain a very large value (at 35° between 100 and 400 atmospheres), this pressure would actually vanish from the equation; whereas it would show a preponderating influence between normal pressure and 100 atmospheres, where it ought itself to explain the *greater part of the variation of volume.*

41

I have already shown that, even after making full allowance for the atomic volume, the occurrence of an internal pressure in such a way as to represent a mere addition to the external pressure is quite insufficient to give an account of the variations from Mariotte's law. The demonstration was on that occasion based on numerical differences of rather small value. To-day, with the new results which I have reached, I am able to make this fact much more evident. I will therefore take up the demonstration again and complete it, introducing in its turn the atomic volume.

Let V be the volume of the gas at the pressure P and the temperature T; let a be the corresponding atomic volume; let the gas be compressed as far as pressure P', and let V' be the new volume at the same temperature. Hence one obtains

$$\frac{(P+p)(V-a)}{(P'+p')(V'-a)}=1, \qquad (1)$$

p and p' being the internal pressures for the volumes V and V'.

Now let the gas be heated as far as T' degrees and let P_1 be the pressure needed to keep the volume at V. The internal pressure, if it depends only on the mean distance apart of the molecules, will again be p.

Compress the gas until its volume has diminished to V' and let P_1' be the pressure needed for this purpose. The internal pressure will again be p' for the reason specified. Hence we should obtain :

$$\frac{(P_1+p)(V-a)}{(P_1'+p')(V'-a)}=1. \qquad (2)$$

Equations (1) and (2) may be put under the following form :

$$(3) \qquad \frac{P(V-a)}{P'(V'-a)}=1+\frac{p'(V'-a)-p(V-a)}{P'(V'-a)};$$

$$(4) \qquad \frac{P_1(V-a)}{P_1'(V'-a)}=1+\frac{p'(V'-a)-p(V-a)}{P_1'(V'-a)}.$$

Put

$$\frac{p'(V'-a)-p(V-a)}{P'(V'-a)}=a, \qquad (5)$$

and

$$\frac{p'(V'-a)-p(V-a)}{P_1'(V'-a)}=a'. \qquad (6)$$

The relations (3) and (4) now become :

$$\frac{P(V-a)}{P'(V'-a)}=1+a,$$

$$\frac{P_1(V-a)}{P_1'(V'-a)}=1+a'.$$

Here a and a' are the variations from Mariotte's law, after the correction for atomic volume has been applied; in other words, the discrepancy of the law:

$$p\,(V-a)=\text{const.}$$

Furthermore, by dividing (5) by (6):

(7)
$$\frac{a}{a'}=\frac{P_1'}{P_1}.$$

Hence the departure from the law $P(V-a)=\text{const.}$ must be in the inverse ratio of the corresponding final pressures P_1 and P_1'.

As an example, data may be given for carbon dioxide at $35°.4$ between the pressure limits 30 and 70 atmospheres. From the isotherms one finds at once that at pressures of 40 and 220 atmospheres at 100° the gas has the same volume as at 30 and 70 atmospheres at 35°.

Hence
$$\frac{a}{a'}=\frac{220}{40}.$$

The volumes V and V' are given in turn by the ordinates corresponding to 30 and 70 meters of mercury. It suffices to divide these by the corresponding pressures. The value of a is deduced from the straight part of the curves of which it is the angular coefficient. Thus without difficulty

$$V=7.9,\ \ V'=1.03,\ \ a=0.625.$$

Hence one should have

$$\frac{(7.9-.625)\,30}{(1.03-.625)\,70}=1+a,$$

$$\frac{(7.9-.625)\,40}{(1.03-.625)\,220}=1+a',$$

whence, after reduction,

$$a=5.65,$$
$$a'=1.81.$$

Finally, in accordance with equation (7),

$$\frac{5.65}{1.81}=\frac{220}{40}\ \text{ or } 3.12=5.5,$$

an absurdity out of all proportion with such discrepancies as might arise out of mere errors of experiment, even when the approximate method of verification is taken into account.

The so-called internal pressure cannot therefore be admitted into gaseous kinetics in so far as this pressure is to depend only on the mean distance apart of the molecules—*i.e.*, to be a

43

function of volume only. It is also a function of temperature, as M. Blaserna* has already inferred elsewhere.

In his calculations of the internal work of a gas M. Hirn makes frequent use of internal pressure. The results at which he thus arrives may therefore appear discordant with my own results. Without wishing to enter into a detailed discussion, I will remark that this disagreement can only be apparent ; it is due simply to the fact that rather in the interior of the molecule than between integrant molecules is the larger part of the internal work expended. It does not follow that the values of internal work numerically calculated are erroneous.

M. Clausius has evolved a theory which has since become classic, and which can very easily give an account of the discrepancies of Mariotte's law. This theory, which in its inception is traceable to Bernoulli, and the first kinetic explanation of Mariotte's law to Krönig, interprets pressure as due to the impact of the molecules of a gas on the walls of the vessel holding it. Pressure thus depends on the kinetic energy of the translational motion.

When a gas is compressed at constant temperature, it suffices to assume that a part of the translational or intermolecular energy is transformed into intramolecular energy or into the energy of molecular rotation, to give a complete account of the discrepancy of Mariotte's law. For the result would be a diminution of pv. As the effect of atomic volume would make the law deviate in a contrary sense, one is led to anticipate the differing phases through which the compressibility of a gas passes, according as one or the other of the two causes supervenes. It is even possible that both causes may annul one another, and that the gas thus accidentally obeys Mariotte's law, as is the case, for instance, in the region of minimum ordinates.

Taken as a whole, the results which I have reached show pretty clearly that a special theory for gases and another for liquids is out of the question. Consider, for example, the isotherms of hydrogen : how is it possible to admit one theory to explain the facts represented by one part of the curves and another theory to explain the rest, seeing that their form shows

---

* Blaserna : *Comptes rendus*, t. lxix., 1869.

conclusively that a phenomenon of perfect continuity has been observed ? Now the theory of impact cannot be applied to hydrogen under 400 atmospheres, since under these conditions the proximity of its molecules would make it rather a liquid than a gas. As long as questions on the condition of carbon dioxide or of ethylene at ordinary temperatures are uppermost, one may infer that the two parts of an isotherm situated on each side of the minimum ordinate are subject to the different laws ; inasmuch as an abrupt variation, which may reasonably be interpreted as limiting two different thermal states, separates them. But this is no longer applicable when the gases are subjected to higher temperatures, nor for hydrogen (as I have stated) even at ordinary temperatures.

It is very remarkable that the law given by the equation $p(V-a)=$const., which has been the immediate outcome of my experiments, and which appears to be the limiting law towards which all gases converge when their temperatures are raised, is the same law which is in action in the neighborhood of the critical point, whenever the compressibility of a body is considered throughout increasing pressures. Thus it is rather a law for the liquid than for the gaseous state. I will even add that it is specifically the law of liquids—at least, within the limits of actual experimental inquiry; for it appears from the group of isothermals that carbon dioxide at 18°, which is then truly liquid, follows exactly the same law. Its curve is a straight line whose angular coefficient is a. Thus we arrive . at a result which appears at first sight quite paradoxical, that elevation of temperature transforms the gas into the state of a perfect liquid; and that the region of branches of the isotherms situated on the left of the line of minimum ordinates, the region which corresponds accurately to the gaseous state in the ordinary sense of the word, is a period of turbulence terminating in the phenomenon of liquefaction properly so called. This phenomenon disappears when temperature rises indefinitely and the body becomes a perfect fluid, to employ a word which is at once applicable both to the liquid and the gaseous states.

The state in question is defined by the simple equation $p(V-a)=$const., or $pW=$const., if W is the interatomic volume. The law thus expressed is so very simple that one is naturally induced to seek an explanation for it depending simply on a consideration of interatomic volume; for what I have already said about internal pressure shows, among other things, that internal pressure does not appear to exert as much influence on the changes of volume of a gas or of a liquid as has been usually supposed. This influence should be reducible to disturbances quite of secondary importance in their bearing on the law in question—disturbances which may, for example, be of the order of magnitude of the discrepancies of Mariotte's law occurring within the limits of pressure explored by Regnault.

To reach an explanation based purely on the consideration of atomic volume, let us observe in the first place that the law $PW=$const. is capable of the following enunciation : The behavior of the body during compression is such as if an infinitely subtle fluid rigorously subject to the law of Mariotte pervaded the whole space between the molecules, the material particles or the groups which they form showing only a negligible amount of translational kinetic energy and producing an effect only by their presence—$i.e.$, by the volume which they delimit in the same way as if they were ordinary walls of the region.

Why may not this fluid be the ether in a certain degree of concentration ? Such an hypothesis would give a complete account of all observed facts. It by no means excludes the theory of molecular impact, as we have seen; it merely restricts the limits within which kinetic action is applicable. Evidently the ether ought to perform some function relative to the phenomenon with which we have been occupied; but this role has never been specified, nor, so far as I know, has anything been said about it. If one considers the exclusive importance of the ether in optic phenomena, the relation of these to thermal phenomena, and the close connection of the latter with those which we have been investigating, the strong probability that the ether must fulfil some important function is manifest.

It seems probable that the molecules are surrounded in every thermal state—solid, liquid, or gaseous—with atmospheres of ether. These atmospheres account for their perfect elasticity, as evidenced in the kinetic theory of gases—an elasticity which it would be very difficult to explain, or which would be even quite inexplicable, if the molecules were simple—$i.e.$, reduced to single atoms. Granting this, let a gas be considered at low pressure and at a temperature but little above the critical point. Let the gas be compressed at an initially constant temperature. The theory of molecular impact deduces Mariotte's law in the usual way. Changes in the distribution of kinetic energy between the motion of molecular translation, the motion within the molecule, and its rotational motion suffice to explain the discrepancies of the law, to which, in a certain measure, molecular attraction may join its effects. Thus the gas is more compressible than Mariotte's law indicates, even if allowance be made for the absolute volume of the atoms. Very soon, however, these with their atmospheres of ether occupy the major part of the volume, and so hamper each other in their movements of translation that the latter virtually vanish. This occurs in the neighborhood of the minimum ordinate. Finally, for continually increasing pressures the atmospheres of ether will actually become contiguous, and the molecules appear as if suspended therein. The ether now forms a medium which is continuous, and by its reaction produces the observed pressure against the walls of the vessel. If this ether obeys Mariotte's law, which is now to be regarded as the limiting law of an infinitely subtle fluid, the volume which it occupies is exactly W, and $PW=$const., or $P(V-a)=$const. Hence, although there has been no liquefaction in the true sense of the word, the body is rather a liquid than a gas, for the reason that the molecular translational motion, which is a criterion for the gaseous state, has vanished.

With the beginning of an increase of temperature, the ethereal molecular atmospheres will expand simultaneously with the molecules themselves, and the atoms separate more and more fully until decomposition ensues. Inasmuch as the total volume of the ethereal atmospheres is larger, the law $pW=$const. ought to begin to apply for a given mass of gas at a larger volume than at the lower initial temperatures. This indeed appears very well to account for the fact that for a given mass of gas the volume corresponding to the minimum ordinate increases with temperature.

If temperature continually rises, the fraction of the total volume occupied by the ether also continually increases, and when the effect of the latter preponderates the curves will rise and be gradually transformed into straight lines.

Perhaps I ought to add, in order to escape an unfavorable issue, that the ether taken into consideration here is that only which is retained by and condensed around the molecules in the form of a molecular atmosphere. The ether which is not so condensed but pervades the molecular spaces, whatever be the distance apart of the molecules, is without relevant influence. This does not oppose any reaction, and for it the walls of the vessel do not exist.

The hypothesis which I have just formulated thus renders a natural and complete account of the details of the phenomena brought out by experiment. It does not exclude those kinetic theories which have gained general acceptance among physicists. So far as I can see, it is not at variance with any established experimental fact. It restricts the limits within which the theory of impact is apparently applicable by establishing a transition from the liquid to the gaseous state, which may be passed continuously and the mechanism of which is easily intelligible. Finally, it introduces no difficulty whatever into the phenomenon of liquefaction properly so called.

I have already stated that the law $PW =$ const., which appears to be the general law for fluids, is applicable to liquid carbon dioxide below the critical point, as is evidenced by the straight isotherm for 18° contained in the family of curves. I endeavored to verify the same fact with several liquids, notably with chlorhydric ether, for which I published the compressibilities as far as 100° and 37 atmospheres several years ago.[*] The law was reproduced with a very fair approach to accuracy; but in order that these verifications may not be over-estimated, it is well to insert the following remark: If the liquid were quite incompressible its isotherm would necessarily be a straight line, for the ordinate would vary proportionally to p, v being constant. For liquids in general, therefore, inasmuch as they are nearly incompressible, their curve must differ exceedingly little from a straight line, no matter what may be the true law of compressibility. It is thus impossible to derive any very certain inferences from the behavior of the greater number of liquids. For liquid carbon dioxide, however, the case is altogether different, and, *a fortiori*, for hydrogen between limits of pressure as far apart as those within which I have operated.

---

[*] *Annales de Chimie et de Physique* (5), t. xi.; *Mémoire sur la Compressibilité des Liquides.*

I have still to give the numerical values of atomic volume, having calculated it for hydrogen, carbon dioxide, and ethylene, all of which contain the straight parts of the isotherms clearly defined and of marked extent. In computing the limiting directions of the lines for the other gases, one is liable to make a considerable error.

To obtain a it is adequate to establish the relation

$$p(V-a)=p'(V'-a)$$

between two pressure values sufficiently far apart and comprehending a part of the curve sensibly straight. Thus the value of a is deduced from p, p', v, v', given by experiment. Hence one obtains the atomic volume of the mass of gas subjected to the experiments from which the curves were constructed. This mass is defined by the normal pressure and the temperature at which the manometers were charged with gas.

I have preferred to refer the atomic volume to the unit of volume of the gas at 0° C. and 76 centimetres of mercury, obtaining the following values :

Hydrogen	0.00078
Carbon dioxide	0.00170
Ethylene	0.00232

Finally, I may remark that the interpretation given for the value of a is independent of the hypothesis on which the theory of the gaseous state is founded. It will readily be seen that it is enough that the curves of compressibility should tend to become straight lines in the. limit; for under these conditions

$$pv=a\,p+b, \text{ or } V=a \text{ for } p=\infty .$$

No matter to what theory one may subscribe, therefore, a appears as the smallest absolute volume which the matter can occupy. One naturally infers that this is the atomic volume. This remark is particularly applicable to hydrogen, the isotherms of which are practically straight throughout their whole extent. For the other gases the determination of a rests on the inferences deduced from the curves as a whole, but less certainly. The results are therefore given with less assurance.

To pursue this subject exhaustively within the limits which I have set for myself, it will still be necessary to study the compressibility under conditions of pressure lower than those

D 49

occurring in the present research. These experiments will be made directly with an open manometer. I hope to carry the measurements as far upward as 300°. I hope also to trace the phenomena further into the region of extremely low pressures —a few millimetres, for instance—on which subject I have already published* an introductory paper.

Returning again to these experiments, I shall be in a position to add many material improvements to the method which I formerly employed, and above all to the apparatus. They will be completed, I hope, in the course of the next academic year.

* *Annales de Chimie et de Physique*, t. viii., 1876

MEMOIR ON THE ELASTICITY AND THERMAL EXPANSION OF FLUIDS THROUGHOUT AN INTERVAL TERMINATING IN VERY HIGH PRESSURES

BY

E. H. AMAGAT

(*Annales de Chimie et de Physique*, 6ᵉ Série, t. **xxix.**, 1893)

CONTENTS

MEMOIR ON THE ELASTICITY AND THERMAL EXPANSION OF FLUIDS THROUGHOUT AN INTERVAL TERMINATING IN VERY HIGH PRESSURES

BY

E. H. AMAGAT

PART I.—METHODS OF MANIPULATION

INTRODUCTION

IN the Memoir which I publish to-day I have brought together the whole of my researches on the expansion and compressibility of fluids,* in so far as they have occupied me during the last ten years. A part only of the results has been published in the *Comptes Rendus de l'Académie des Sciences;* the experimental portions, moreover, were sketched with the utmost brevity compatible with clearness.

The present researches for gases are a direct continuation of my earlier work on the same subject. In the latter the limits of pressure and of temperature employed were too narrow, and the number of isotherms mapped out not great enough to reveal certain relations which appear very clearly in the present results. Such are, for example, the form of the isotherms in the region of the critical point which lay beyond the limits of my first group of curves; furthermore, the contours of these

---

* The parts of this great memoir referring to liquids will not be included in the present translation.

curves in the region lying to the right of the locus of minimum ordinates, and in which the isotherms under very great pressure seem to merge into straight lines, etc.

As for liquids, the question was virtually untouched when these researches were begun. The increase of the coefficient of compressibility with temperature had been observed for some liquids, together with the contrary effect for water. I myself extended these results, in a memoir published in 1877, to a large number of liquids, and within limits of pressure and of temperature beyond any which had been applied at that time; but the laws as a whole were yet to be investigated, the determination of pressure coefficients was not even attempted, the data relative to the variation of the coefficient of compressibility with pressure were altogether contradictory. Well-known treatises of physics even to-day contain errors in relation to this subject which surpass the limits of plausibility. Since that time, however, several important researches bearing on these phenomena have been published outside of France.

The researches which I am about to describe have been made with a view towards reaching the highest attainable pressures, both for liquids and for gases. It is my purpose, furthermore, to make special investigations for the low pressures—*i.e.*, for the first one hundred or two hundred atmospheres, and thereafter to co-ordinate them in such a manner as to give a complete presentation of the phenomena. Experiments devised to reach a thousand or several thousand atmospheres must at lower pressures of necessity show an inferior degree of accuracy than may be reached in experiments specially adapted to the latter. Circumstances have not permitted me to terminate this research, but the most difficult part of it is finished.

In this place I may be permitted to say that the instruments which are to be described, as well as all others used in my researches for upwards of fourteen years (1877 to 1891), were constructed, in the workshop attached to my department (*service*) at Lyons, by M. Gianotti (at present instrument-maker in Lyons). My tasks have, throughout, been singularly facilitated both by his skill in construction and by the earnestness with which he aided me in the experimental work.

Glass apparatus was made by M. Alvergniat, or by his successor, M. Chabaud. Indeed, all the apparatus, either of ordinary or of cut glass, which has been used in my experiments for twenty-five years or more, was constructed by these gentlemen. I need not further advert to their services, cheerfully rendered in the cause of science.

I shall first describe the methods and the apparatus. They are of like construction, both for liquids and for gases, except as to the piezometer containing the fluid and the manner of charging it. I shall begin with the apparatus for pressure measurement.

METHODS OF EXPERIMENTATION

The Measurement of High Pressures—" Manomètre à Pistons Libres."

.

When I undertook the present researches there was no instrument available for the accurate measurement of pressure beyond the range within which it was customary to compare the well-known empiric pressure-gauges with the open manometers. In practice closed gas manometers are subject to serious inconveniences. Moreover, they cannot be employed above 420 atmospheres, this being the upper limit of the measurements which I made, in 1878, with an open manometer in the shaft of the mine at Verpilleux.

The principle of the instrument improperly called *manomètre de Desgoffe** solves the question theoretically; but the practical construction adopted was actually so defective that no reliance could be put on its indications. The manometers of the valve or plug type may perhaps render service in certain cases. M. Marcel Deprez has improved them by replacing—for the first time, I believe—the valve by a free plunger (*piston libre*), which does not allow water to pass except with extreme slowness. But these instruments even when perfected are not available in researches which cannot well be conducted without a pressure-gauge of continuous registry.

The grave difficulty in the way of a realization of Gally-Cazalat's idea is this : to make the pistons perfectly free to move while at the same time obviating leakage. For the large piston the difficulty is in a certain measure solved by the ad-

* This instrument was invented by Gally-Cazalat ; constructed at first by Clair, then by Bianchi, and finally by Desgoffe.

dition of an india-rubber membrane. Nevertheless, this ingenious device is not quite beyond criticism ; for in the first place the sectional surface is badly determinable, while in the second the action of the membrane gives rise to an error which the operator must either endeavor to estimate or to obviate. In the first instruments which I constructed I retained the membrane design, but an index rigidly attached to the large piston enabled me to follow its motion, and, therefore, that of the membrane also. It was then my purpose to provide the apparatus with a regulating pump, which by injecting a variable quantity of liquid below the membrane would enable me to keep it always in the same horizontal plane, thus suppressing nearly completely the action referred to. Later I found it more advantageous to do away with the membrane altogether and to leave the large piston entirely free.

To secure freedom from leakage, I found it sufficient to give the piston a suitable thickness and to replace the actuating water by a lubricating and at the same time slightly viscous liquid ; castor-oil is very serviceable for this purpose.

The analogous difficulty remained for the case of the small piston, which, experiencing strong pressure in a leather-lined stuffing-box, moved only with difficulty and by jerks. It was necessary to make this piston free, like the first, but the condition of no leakage was here very much more difficult of attainment in spite of the small section. For while the large piston receives only the relatively small pressure of the counterpoising column of mercury, the enormous total pressure which is to be measured bears down upon the small piston. I succeeded, however, in meeting the present difficulty by the identical artifice—i.e., by using a sufficiently viscous liquid, in this case molasses. Nevertheless, the function of this body is somewhat different from that of the castor-oil : for while the oil continually lubricates parts, oozing with extreme slowness between them in a way not to interfere with effective action, the molasses penetrates the space around the piston—supposed to be well oiled in advance—with very great difficulty even at very high pressures. When the molasses has succeeded in penetrating and removing the oil, the apparatus still functions, although with loss of some of its original sensitiveness. It is then preferable to detach the small piston and to clean it. When the apparatus has been adjusted with care, it is sufficient to place an object of even insignificant weight on the large piston to produce a corresponding small ascent of the column of mercury. The small piston is in general less sensitive, above all after the molasses has penetrated between it and the socket. Finally I succeeded in quite annulling the resistance due to friction by impressing on both pistons a slight movement of rotation. An analogous artifice has long been employed by

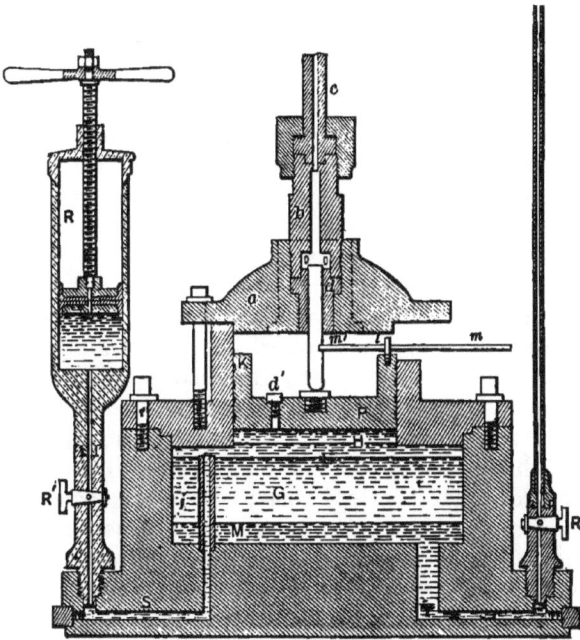

FIG. 1.—FREE-PISTON MANOMETER (*Manomètre à Pistons Libres*).

M. Bourdon to overcome the friction of the piston of the apparatus used in standardizing his spirals; but while the leather-lined stuffing-box requires a rapid movement of rotation in order to obviate the friction of a tight fit, my apparatus needs only a slow and slight angular displacement of the pistons in order that the column of mercury may at once take its definite position of equilibrium.

Fig. 1 gives a section of the apparatus. The liquid transmitting pressure arrives by the tube, *c*, through a channel in the piece of steel, b, screwed to the brass * lid, a. This piece secures the socket of tempered steel, d, holding it down free from leakage by aid of round leather washers. The small piston, also of tempered steel, moves up and down in this socket. The parts are so fashioned that a small chamber, OO, is left below b. Into this the charge of molasses is to be put. The lower end of the small piston abuts against a small plane of tempered steel, which is seen in the figure, screwed to the centre of the large piston, P. A valve-screw, d', may be raised to admit of the escape of air from below in adjusting the apparatus. Circular grooves are cut equatorially around the outer walls of the large piston. The oil accumulates in these channels during its upward leakage, and finally reaches the hollow part or cup of the large piston. The latter moves up and down in a massive envelope, also of brass. This is screwed down to a heavy trough of cast-iron, serving as the base of the apparatus, by a crown of square-headed bolts, and further secured to the lid, a, of the apparatus by a second crown of longer bolts. A key, not shown in the figure, may be attached to the large piston at its centre, for the purpose of withdrawing or of inserting it. In such a case the key replaces the small plane of steel.

On the right side of the figure, and screwed to a lateral projection of the trough, is the steel coupling carrying the glass tube in which the mercury column rises. This coupling consists of two parts: the lower part carries a stopcock, and thus the operator is able to remove the glass tube, even when the trough is charged, without spilling the mercury within it. At the left of the figure, and symmetrically placed, is the regulating pump, and this, for a reason similar to the one just given, also consists of separable parts with a stopcock in the lower. The channel by which the oil, H, is injected is prolonged by a tubulure extending upward much above the mercury surface, M, in order that this may under no condition reach the pump, which is of brass, and thus in danger of amalgamation.

The angular movement of the two pistons is produced by the steel rod, $m\,m'$, screwed † to the prolongation of the small pis-

---

* In the case of instruments for measuring very high pressures, the strength of the brass bolt crown is insufficient ; the central part of the lid, as shown by the dotted lines, should be made of steel.

† For pistons of very small diameter, this prolongation is enlarged and different in form.

ton. This rod passes between two pins, i, screwed to the margin of the large piston, and leaves the apparatus through a window in the upper part of the cylinder. Thus both pistons may be put into the same horizontal angular motion, which is transmitted by a simple arrangement not shown in the figure. The regulating pump makes it possible to keep the stem, $m\,m'$, at a height such as will not interfere with the upper or lower edge of the window. Moreover, by an inverted action of the appurtenances of the manometer, the pump may be made to produce very considerable pressures above the small piston and in the space where the bodies subjected to experiment are placed; it is often convenient to make use of this method to regulate the pressure at the moment of measurement. M. Vieille * has recently made a very happy application of this artifice, in connection with his studies on the behavior of the crushing manometer (*manomètre-crushers*).

If well constructed, my apparatus will give pressure-values of remarkable uniformity, the accuracy of which depends only on the sectional ratio of the two pistons. The sensitiveness can be increased at pleasure by increasing the value of this ratio and making use of a graded series of pistons. The apparatus which I used was provided with two large pistons, the one 6 centimetres and the other about 12 centimetres in diameter, together with a series of small pistons, the smallest being 5.527 millimetres in diameter.

The absolute error is about the same throughout the whole scale of pressures. The relative error would become intolerably large if a few atmospheres only were to be measured, which, however, is beyond the purposes of the instrument. It is, nevertheless, my intention to construct a small model specially designed for the lower order of pressures, and I hope to find it relatively quite as reliable.

On several occasions I made comparisons of data for the same pressures furnished simultaneously by two different free-piston manometers, or when one of these was replaced by a closed gas manometer. The agreement of the former was always very satisfactory, but this was usually less so with the gas manometers. I do not hesitate to ascrihe these discrepancies to the difficulty of manipulating the latter. Here is the com-

* *Mémorial des Poudres et Salpêtres*, v.

parison of the free-piston manometer which I constructed in 1885 for M. Tait, with two nitrogen manometers and an air manometer. Pressures are given in atmospheres.

TABLE 1.—MANOMETERS

NITROGEN	PISTON	NITROGEN	PISTON	AIR	PISTON
226 Atm.	224 Atm.	102 Atm.	103 Atm.	217 Atm.	215 Atm.
278 "	275 "	154 "	154 "	262 "	263 "
328 "	326 "	213 "	215 "	305 "	306 "
391 "	387 "	256 "	257 "	358 "	362 "
438 "	438 "	299 "	297 "	401 "	406 "
		363 "	359 "		
		408 "	402 "		

Never did the comparisons of free-piston manometers present like divergencies.

The following table contains a comparison between the results given by a membrane manometer with but one free piston and a manometer with two free pistons :

TABLE 2.—MANOMETERS.

MEMBRANE.	TWO PISTONS.	RATIO.
103 Atm.	102 Atm.	1.010
156 "	154 "	1.013
215 "	213 "	1.010
260 "	257 "	1.012
304 "	301 "	1.010
368 "	364 "	1.011
444 "	439 "	1.011
550 "	545 "	1.009
604 "	600 "	1.007
697 "	692 "	1.007

The agreement of piston and gas manometers improves when the latter are used under the best conditions.

In Table 3 are the results of a comparison with an air and a nitrogen manometer, devised for very high pressures, much more sensitive than the above, and directly graduated for each gas by comparison with a column of mercury in one of the towers of the Fourvière cathedral in Lyons—that is, under the best conditions possible. The free-piston manometer was itself so adjusted that fractions of an atmosphere were easily read off.

TABLE 3.—MANOMETERS

AIR	PISTONS	NITROGEN	PISTONS	OPEN MANOMETER	PISTONS
26.36 Atm.	26.32 Atm.	26.29 Atm.	26.50 Atm.	1.66 Atm.	1.74 Atm.
32.39 "	32.34 "	32.51 "	33.59 "	2.95 "	3.02 "
38.34 "	38.44 "	39.12 "	39.21 "	5.07 "	5.16 "
44.98 "	45.00 "	45.77 "	45.81 "	6.52 "	6.61 "
50.92 "	51.05 "	52.26 "	52.41 "		
57.37 "	57.50 "	58.12 "	58.87 "		
64.24 "	64.16· "	65.35 "	65.53 ··		
72.15 "	72.45 "	71.00 "	71.36 "		

Finally, in the last two columns, the table contains a comparison, throughout a few atmospheres only, with an open mercury manometer, the errors of which may be regarded practically zero. The discrepancy is obviously inadmissible for pressures as small as these; but for higher pressures, 50 or 60 atmospheres for example, the absolute error not becoming larger, the results are exceedingly satisfactory.

For the series of measurements reading upward as far as 3000 atmospheres, I chose a pair of pistons such that one atmosphere was equivalent to 1.601 millimetres of mercury at 0° C. For the series limited by 1000 atmospheres, the equivalent height was 4.99 millimetres. In each case I selected those sectional ratios which gave me the highest attainable sensitiveness for the case of a column of mercury 5.20 metres in height, the maximum elevation at my disposal.

2. APPARATUS FOR EXTREMELY HIGH PRESSURES

Method of Electrical Contact.

The methods of which I made use in my preceding researches do not admit of being carried much above 400 atmospheres. It is very difficult to obtain glass tubes which will resist interior pressures more intense than this. To solve this difficulty by plunging the piezometer bodily into a stronger

• cylinder, necessarily metallic and opaque, presents grave difficulties in regard to the reading of volumes. Otherwise it is the method of Oersted. The difficulty in question was first avoided by means of gravitational apparatus, or by covering the interior of the stem of the piezometer by a film which is dissolved by the liquid transmitting the pressure, thus indicating the height to which it has been raised. These procedures have two serious defects : aside from the fact that they merely give the maximum of the height to which the liquid has been raised, they are quite impracticable for the construction of a regular series. They have given rise to very grave errors. The method pursued for gases by Natterer, which consisted in compressing a known volume of gas into a given space, repeating the operation a great number of times and determining the pressure corresponding to each accession, is certain to be surrounded with great difficulties, and rapidly becomes quite impracticable if one endeavors to raise the temperature. The work of Natterer is none the less extremely remarkable, particularly if the time when it was done (1851) is called to mind. It is hard to explain why it has remained so long unknown in France.

At the time when I began these researches—that is, about 1882—M. Tait was engaged with a study of the compressibility of water, and he called my attention to the method of electric contacts which he employed in his researches. Among other things, I had also thought of this artifice, but I had as yet made no trial of it. Upon the recommendation of the eminent physicist I tried it at once ; and since then I have employed no other method, either for liquids or for gases, in all series of measurements carried to the highest attainable pressures and at temperatures not exceeding 50°.

Fig. 2 gives a sectional elevation of the apparatus constructed for these researches. The receiver, or barrel, GG, in which the piezometer is placed is a cylinder of steel 3 centimetres in diameter internally, and surrounded by a steel jacket, $G'G'$, to a point about as far down as the bottom of the aperture within. The total outer diameter is 18 centimetres, the available depth of the receptacle below the position of the upper junction is about 88 centimetres. The somewhat excessive prolongation of the unjacketed breech has a special purpose to which I shall refer below. Pressure is transmitted by a charge

of water injected by a force-pump, and enters at D by way of
the valve-coupling, E, screwed to the upper part of the receiver
on the right. When the pressure attains a certain value de-

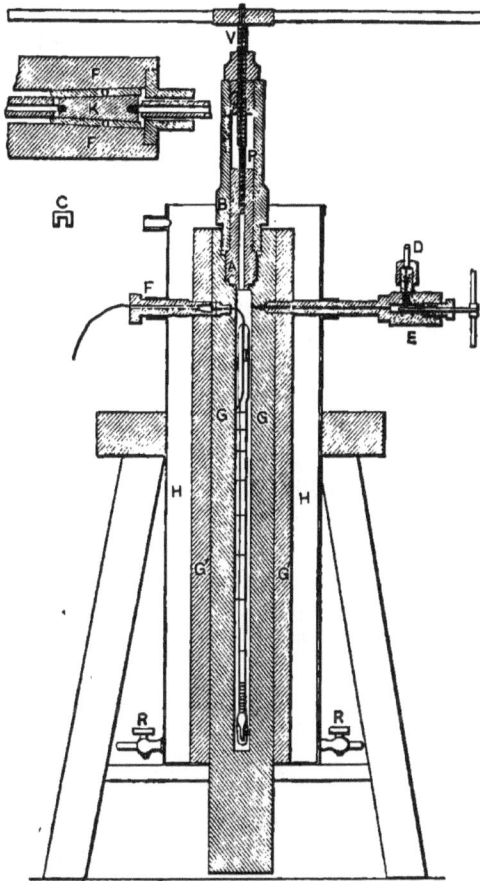

FIG. 3.   FIG. 2.   FIG. 4.

FIG. 2.—PRESSURE APPARATUS WITH ELECTRICAL CONTACTS.
FIG. 3.—PIEZOMETER FOR GASES.
FIG. 4.—PIEZOMETER FOR LIQUIDS.

pending on the upper limit which is to be reached, the screw-valve is closed, and compression is thereafter continued with the use of the device screwed to the head of the receiver. The central piece, A, is pierced by an aperture 16 millimetres in diameter, in which a cap-shaped leather washer, C (drawn in larger scale at the side), moves up and down. This washer has the form of a little cylinder with a flat base. It is pushed forward by a rod of steel, P, advancing with slight friction and terminating in a cone at its upper end. Here the motion is impressed upon it by a steel screw, V, the threads of which are 2 millimetres apart and move in a nut of brass carried by a second hollow cylinder of steel, B, surrounding the central piece, A. This second cylinder is screwed to the jacketed receiver, against which it presses the central piece, A, making a pressure-tight joint with it by the aid of a leather washer. The screw is actuated by a four-armed lever seen at the top of the apparatus. At the left, on the same level with the influx valve, there is a special attachment, F, of steel, adapted to carry the electric current into the piezometer, insulated from the body of the apparatus. This is attained by aid of a small cone of steel, K, preliminarily inserted into the channel within the piece, F, and which is wedged after having been surrounded by a very thin, small conical shell of ivory, OO, insulating perfectly. Moreover, the taper of the cone is so directed that the internal pressure will force it into more intimate contact with its surroundings. The small steel cone joins the ends of the conducting wires, which are screwed into it in the manner detailed in the enlarged figure. This attachment has never shown the least trace of leakage.

On a level with the influx valve and the attachment, F, there is still a third coupling, not shown in the figure, which puts the barrel in connection with the manometer (these three pieces are in reality placed so as to divide the circumference of the cylinder into three equal parts). The connecting tube is about 2 meters long and made up of three parts bored at the lathe, the last piece being afterwards bent down by forging. I at first employed thick tubes of drawn steel, but they burst at about 2000 atmospheres.

The piezometer containing the fluid to be compressed is of a form shown separately in Fig. 3 for gases and in Fig. 4 for liquids. The stem carries a series of fine platinum wires laterally inserted into the glass and reaching as far as the middle of the bore. Between each thread there is inserted an electric resistance wrapped around the tube. These resistances are made up of wire covered with india-rubber, and a small part of each coil is exposed so as to be soldered to the corresponding ends of the platinum wires. The whole is then covered by a wrapper, insulating perfectly even when plunged in mercury, and remaining sufficiently soft to insure an equal compression of the glass on opposite faces. When the piezometer is placed within the barrel, the upper terminal is joined to the prolongation seen within the cone of the attachment, *F*, contact being thus completed throughout. When the mercury into which the lower end of the piezometer is plunged rises in the tube conformably with the increasing pressure, and just touches the first of the lateral platinum wires, the current may be closed between any part of the barrel or its metallic appurtenances and the conductor joined at the outer end of *F*.

The jacketed part of the barrel is further enveloped by a spacious brass cylinder filled with ice or charged with water kept in circulation and issuing from special auxiliary apparatus adapted to secure constancy of temperature between 0° and about 50° C. Entering near the bottom, the circulating water issues from a tubulure near the top. A thermometer shows its temperature. Two stopcocks, seen at the bottom of the water-jacket, facilitate influx. The bath is surrounded with wood sawdust, contained between the four braced timbers of oak, and kept in place by shutters not shown in the figure. The upper part is protected with felting.

Details of Manipulation.—Let the case of a liquid be first considered. The piezometer, Fig. 4, clean and dry, is filled with the liquid, which must have been boiled to remove the air in the usual way. This done, a column of mercury several centimetres high is introduced at the lower end of the stem by aid of a funnel, the end of which has been drawn out to as long and fine a point as desirable. This end of the stem is then plunged into a small glass full of mercury, and the body of the piezometer carefully heated until the meniscus reaches the lower end. On cooling the mercury again rises, and the column so formed is very uniform and free from breaks due to the other liquid. Thereafter the stem is provided with a small cylindrical reservoir of steel, and the stem dips into the

E 65

mercury contained as shown in the figure. It is now necessary to find the mass of the liquid operated on. For this purpose the piezometer is submerged in a long test-tube of glass containing mercury at its lower end, into which the little steel reservoir is plunged. A current of water at constant temperature is made to circulate in the test-tube, and when a state of thermal equilibrium has been reached the volume of the liquid is read off on a standardized scale etched into the lower part of the stem. The piezometer may now be put into the barrel of the pressure apparatus, also charged at its bottom with a sufficient quantity of mercury. The electric circuit is then completed, a spring-stop added to hold the piezometer in place, the attachments for the production of pressure screwed on, and all is now ready for making a series of measurements.

In the case of gases, the following operations are necessary: The current of gas is passed into the piezometer, perfectly dry, and heated from time to time during the operation, and passed out through the fine point at the top. At suitable periods a sample of the gas is tested, and when the examination shows that the piezometer contains perfectly pure gas only, the point at the top is closed with the blow-pipe. Thereafter, without separating the piezometer from the purifying and drying train in which a small excess of pressure is kept in action, the piezometer is received and held vertically by a tube of brass in a special apparatus, which in turn is submerged in a glass trough supplied with a current of circulating water. When the temperature has become stationary, the normal pressure is again established in the dissicating trains, and the rubber tube connecting this apparatus with the lower end of the piezometer, which alone is outside of the bath, is removed with caution. At the same time this end is plunged into a cistern of mercury. This operation to be well made requires some skill and practice. It is now only necessary to fit the small steel thimble to the lower end under mercury, and to place the piezometer in the barrel. All heating which may cause the gas to flow out during the adjustment is to be scrupulously avoided. It is obvious that the pressure of the barometer was taken at the required time, and that the mass of gas is thus perfectly determinate.

All piezometers, whether adapted for liquids or for gases, were calibrated with mercury in such a way that the instant at which the mercury touches one of the lateral platinum filaments of the tube is shown electrically precisely in the manner to be adopted during the experiments. The volumes given by the calibration tables are therefore quite identical with the corresponding measurements under pressure.

For the gas piezometers, in particular, this standardization is a rather delicate operation because of the small volumes to be measured. I carried it out in two ways, directly in weight and indirectly in volume. By aid of a standard tube calibrated with great care and temporarily soldered to the stem carrying the platinum contacts, I measured the volume of mercury which is withdrawn from this stem while the column is made to glide from one contact to the next. In this case the small olive-shaped reservoir just below the upper point is calibrated separately. The point, though very fine, is itself calibrated from a fiducial mark onward, in order that the diminution of volume produced when the fine point is cut off and resoldered may be estimated, however small it may be.

Plan of a Series of Measurements

Fig. 5 shows the completed apparatus and its accessories. In the rear of the figure is the tank containing the water to be heated. Constancy of temperature is obtained by the aid of a crown burner of gas, a regulator, an influx tap, and an overflow, all conveniently disposed. For the lower temperatures, water cooled down as far as zero in a refrigerator is raised to any desired degree of temperature by its passage through longer or shorter spirals kept within appropriate temperature baths. On the left is seen the large force-pump communicating with the corresponding stopcock and coupling. Quite in front is the free-piston manometer with the lower end of the mercury column. At the right, on a bracket, is the galvanometer, inserted into the electric circuit to announce the galvanic contacts. It is seen at once how the first instant of contact is obtained. For the others I adopted the following arrangement : The current is branched and the galvanometer mounted differentially. One of the shunts passes through the resistances of the piezometer, a box of resistances, and a rheostat. The galvanometer is put back to zero after each contact, and these are unmistakably indicated by the suppression of a definite quantity of piezometer resistance whenever the contact is made. The force-pump is first to be actuated in order to completely fill the apparatus and to start the excess of internal

Fig. 5.—Disposition of Apparatus for Very High Pressures

pressure. To find the exact instant of contact use is made of the compressing screw, which, moreover, is employed exclusively as soon as the pressures reach values of 400 to 500 atmospheres.

When a contact is obtained it is necessary that the apparatus should return again to its initial temperature, since this has been changed by the thermal effect of compression. The initial conditions may be considered as established when, on breaking and reproducing the contact many times by means of small pressure increments slowly applied, it is found to recur uniformly at the same pressure. This pressure is then marked with a fine chalk pencil on the scale of the mercurial column. The galvanometer is now set back to zero, while the compression is continued as far as the next contact, progressing in like manner until the last contact is reached. Thereafter the series is repeated throughout in the opposite direction—*i.e.*, all contacts are passed again in a march of decreasing pressures.

If the series has been carried to completion rather too rapidly, it will happen, owing to the inverse thermal effects during the periods of increasing and decreasing pressures, that the pressures observed in descending will be a little smaller, *caet. par.*, than those observed in ascending. The difference is usually very small, and the mean may be taken. Its value at the same time is a criterion of the degree of accuracy guaranteed.

When the constancy of temperature in the apparatus has been exceptionally good, and if the observer has waited long enough at each contact, the pressures in the ascending series are exactly reproduced by the corresponding pressures in the descending series. This is the best proof which can be given of the trustworthiness of the instrument used in measuring pressures.

Work may be done more expeditiously by completely submerging the piezometer in mercury. The thermal effect of compression is then small, and by reason of the good condition of mercury, the thermal equilibrium is re-established very quickly. In this case, however, the insulation of the resistances requires much greater care.

When the cylinder and the piezometer are filled with water it is perfectly feasible to put in evidence the reversal of the thermal effect of compression on the respective sides of the temperature of maximum density of water.

These operations, which require the concerted observation of at least three persons, are long and tedious. A single series may extend over two, three, four hours, or even more, depending on the number of contacts and supposing that no accident occurs.

I stated above that the prolongation of the unjacketed breech of the steel cylinder seemed to be of excessive length. The reason of this design may be found in an accident which happened to me in the experiments in which oxygen was compressed as far as a maximum density above 1.25 (relative to water) at the temperature of 17° C. The cylinder was quite strong and filled with mercury.* Suddenly, with a strident noise, a jet of pulverulent mercury was hurled across the right section of the breech, striking the base of the apparatus, rebounding thence to more than a metre in height in all directions. The noise was like that produced by a jet of steam escaping from a boiler under high pressure. A right section of the column after being polished showed nothing in particular when examined under the microscope. We have thus encountered the classical experiment of the rain of mercury through the pores of a body of steel about .08 metre thick. The pressure was certainly as high as 4000 atmospheres. The same apparatus under the same pressure did not admit of the exudation of a single drop of glycerine. It would doubtless have shown the same negative result for water and for other liquids.

No similar accident occurred during the course of my work, so far as a flow normal to the section of the cylinders used is concerned. The reason why the thickness of the breech was increased to an extreme degree in the direction of the axis is, therefore, clear.

I was induced to devise a jacketed cylinder in consequence of an accident of another kind. The first large cylinder of steel which I used (weighing 116 kg.) split apart along opposite generatrices throughout a considerable part of its length. Although there was neither projection nor separation of parts, nor any sudden issue of gas, the rupture was nevertheless accompanied with a detonation of extreme violence. During a few moments the mercury escaped from the fissure, and I had time to observe it in form of a bright metallic plate 6 to 7 cen-

* *Comptes Rendus*, March 2, 1885.

timetres in breadth. It was owing to the occurrence of these two accidents that I added the steel jacket to the cylinder, as already described.

3.—APPARATUS FOR HIGHER TEMPERATURES
Method of Sights.

It would be difficult to work at temperatures markedly higher than those for which the apparatus just described was designed. The mass of the jacketed steel barrel, the presence of the appliances added at its top and which project outside of the bath, would make a uniform degree of high temperature very difficult of attainment. The joints and the cap-shaped washer would no longer insure freedom from leakage. Finally and particularly, for the case of gases, the piezometer stems, fragile at best by reason of the inserted platinum filaments, would become prohibitively so, while the difficulties of an adequate insulation of the parts would in like measure greatly increase.

The following design, which I shall call the method of sights (*méthode des regards*), has enabled me to work as far as 360° C. It would even be possible to reach higher temperatures, in modifying the apparatus in the way which experience has suggested, but which I do not expect to apply at present. I have also thought it desirable to restrict the pressures within 1000 atmospheres, although in many trials I went much beyond this.

The method may expediently be described together with the apparatus. The latter is represented in sectional elevation in Fig. 6. In the lower part of the apparatus the jacketed barrel, *HH*, of the preceding compressor will be recognized, although three breaks were needed to shorten the figure. At the top of this is fixed (as above) the apparatus for producing pressure, modified by the introduction of a cross of steel, *A, A, F, F,* forged and thereafter turned and pierced at the lathe throughout the axis of the arms. The horizontal arm, *AA,* carries the bolts, *BB,* screwed to its two extremities, suitably tubulated, in which the sights are cemented with marine glue. These are small cylinders of crown glass or of quartz, with good plane parallel faces, about 1 centimetre in diameter and 2 centimetres long. The joint with the gland is

71

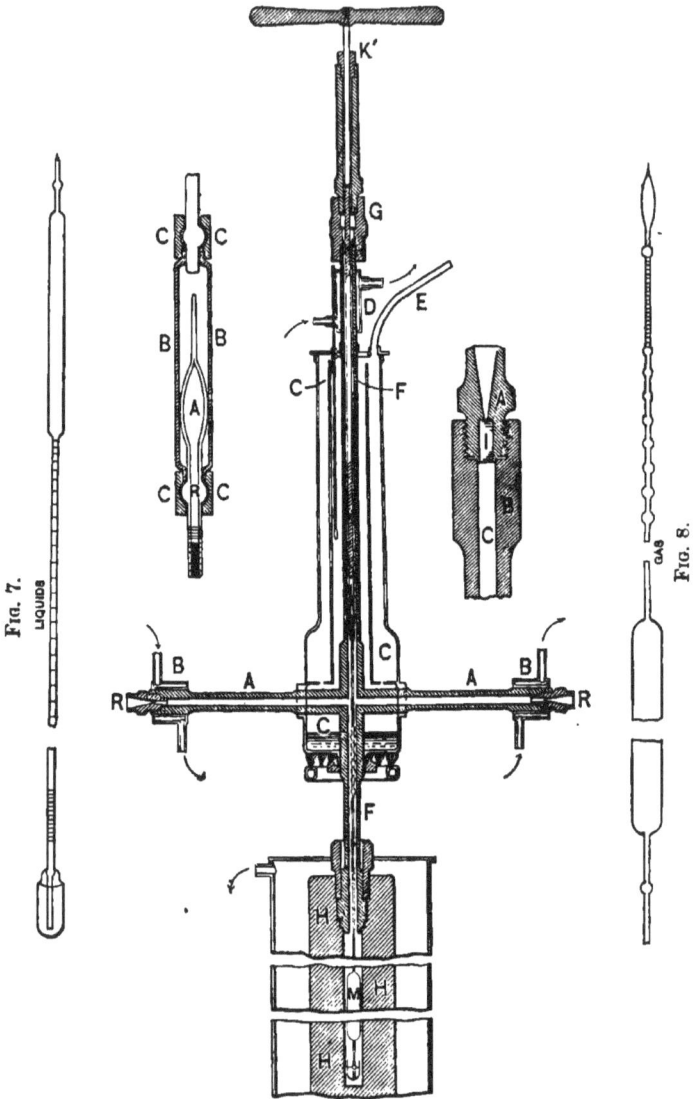

FIG. 7.

LIQUID

FIG. 8.

GAS

C   C

B   B

A

C   C

C   C

K'

G

D   E

C   F

A

I

B

C

B   B

A   A

R   R

C

C

F

H

M H

H U

FIG. 6.

FIG. 6.—APPARATUS FOR THE METHOD OF SIGHTS.
FIG. 7.—PIEZOMETER FOR LIQUIDS.
FIG. 8.—PIEZOMETER FOR GASES.

sealed with a washer of celluloid. The piezometer is mounted symmetrically with the axis of the apparatus, and the figure shows a piezometer of gas in place. Its reservoir is submerged in mercury, which now partially fills the barrel. Its very long stem terminates above in a small olive-shaped reservoir ending in a fine point, as above. Below the enlargement is the graduation, about 25 centimetres long and made of very delicate circular marks narrowing towards the top. Little globular expansions were blown out between them along the lower part of the stem, with the object of virtually increasing the total length of the graduated part. The piezometer is suspended from its upper end by aid of a small pad which is attached to it under the uppermost bulb, and sustained by the aid of a device, *BB*, *CC*, shown enlarged in a separate figure. This piece is connected at *CC* to an intermediate stem of glass joined in the same way at its upper extremity to a long rod of steel. The latter, after having traversed a stuffing-box charged with leather washers, is in its turn joined to the lower end of a long steel screw. This is movable in a socket of brass carried by the same coupling which receives the leather of the stuffing-box. The jacketed barrel, part of the barrel *HH*, is as usual provided with a stopcock (not shown), through which the initial pressures are applied in virtue of the force-pump. Another tube communicates with the manometer. In place of the third coupling, which, in the former case, supplied the electric current, there is now a tube of steel leading to a special steel reservoir, to which the device for producing pressure, formerly described, is suitably attached. High pressure is, therefore, brought to bear at this place by manipulating the lower screw.

The method of experimentation will now be intelligible without much further explanation. By means of the screw at the top of the apparatus, the division-rings on the stem of the piezometer are successively placed in the line of sight. Thereupon pressure is applied until the meniscus appears flush with the division mark, and this pressure is registered. The readings are made with a reading telescope well centred in the line of sight, and the field of vision is illuminated with a simple gas-lamp placed on the opposite side in the same direction. This light is quite sufficient as long as certain precautions are taken which I will now indicate. The injection water rapidly loses its transparency, particularly when the stem is heated. Reading thus becomes more impossible in proportion as the layer of liquid penetrated by the rays is thicker. To obviate this difficulty I at first placed cylinders of crown-glass, plane-parallel and perfectly transparent, throughout the whole length of the channel. Readings at the lower temperatures were then easily made, but at the higher temperatures the faces of the glass cylinder were rapidly attacked in the hot regions and after a time covered over with an opaque pulverulent layer, being thus completely corroded. Hence, in subsequent work I replaced the crown-glass cylinders in the hot parts of the tube by quartz cylinders with their faces normal to the axis. Readings could then be made without difficulty. Again, to avoid the decomposition of leather, which would have clouded the internal faces of the sight-cylinders, the joints in the nuts and valves were sealed with celluloid washers.

The different temperatures at which it was proposed to investigate were obtained by enveloping the arms of the cross, AF, by an appliance adapted to do service either as a water-bath or as a vapor-bath. The lower part of this surrounded the three arms permanently, the seal being made with red-lead. The upper part of the bath differed in form according to the uses to be made of it, and it was fitted into a shouldered rim in the lower part and held in place by friction.

To obtain a vapor-bath an appliance with a two-chambered interior was at hand, provided with a perforated bottom. The condenser communicated with the tubulure F of the removable lid, which at the same time furnished a support for the thermometers. When a liquid bath was wanted, the partition was replaced by an agitator actuated by a small Gramme machine. Constancy of temperature was then obtained by suitably adjusting the distance and the gas supply of the crown burner seen at the bottom of the bath.

At D, under the leather stuffing-box, and at the extremities, BB, of the horizontal arms, small water-jackets were added, fed with a current of cold water. This prevented serious overheating of the leather washers or of the mastic seals at the sights. The discharge water flows into the lower reservoir, keeping the joints of the cross, which are plunged in it, cold, and then escapes by a lateral tubulure.

The result of this cooling is that under the stuffing-box, G. in particular, the cross reaches the temperature of the bath only at a considerable distance below the lid. I thought it worth while to make direct tests by using long, probelike thermometers specially made for this purpose, in order to be properly guided as to the maximum height up to which the tip of the piezometer could be raised at any temperature without encountering seriously reduced values. It would have been far preferable—indeed, at higher temperatures, it would be absolutely necessary to close the cross immediately below the stuffing-box, and to produce the vertical displacement of the piezometer by means of an appropriate appliance attached at the lower end. All this is feasible, though not without difficulty.

PRELIMINARY OPERATIONS

After what has been said relatively to the first method, only a few words are needed. The piezometers, Figs. 7 and 8, both for liquids and for gases, do not differ from those above described except as to the stems, which now carry the circular division marks in place of the platinum filaments formerly used. They are filled and the mass of the contents determined in identically the same way as before, but it is much more difficult to adjust them in place. This can only be done by a suitable pulley-block fastened to the ceiling, by means of which the cross may be raised or lowered without the least jolting. Even slight percussion would invariably break the stem of the piezometer.

Constancy of temperature is now reached much more rapidly than in the preceding experiments, whether the environment be an ice-bath, a water or a vapor bath; for the total mass which is to reach a stationary temperature distribution is enormously smaller. A full series of experiments is nevertheless tediously prolonged, seeing that the number of division marks on the stem is so much larger than the number of platinum filaments above. As far as 100° the temperatures were given by water-baths. A point very near 200° corresponds to the boiling-point of methyl benzoate, another at 260° to amyl benzoate. Many bodies were tested as to their availability in vapor-baths between 100° and 200°. For 140° I employed xylene and ethyl acetate, but the results were less satisfactory than in the preceding cases. I was quite unable to obtain a perfectly constant temperature from xylene. The specific heat per unit of volume seems to be very small, and the heat transferred insufficient to compensate for the external losses. I shall therefore publish the series made at this temperature with this special reservation.

In all the series of measurements made up to 100° by this method, I restricted the observations to an exact number of degrees, using thermometers for this purpose compared in advance with a standard provided by the *Bureau International des Poids et Mesures*, and calibrated by M. Guillaume. They are reduced to the hydrogen thermometer by means of the table of M. Chappuis. At 100° either a water-bath or a steam-bath was available. I have given the results for 100° exactly, the interpolation being insignificant and no error resulting from it. Temperatures above 100° were determined by means of excellent thermometers of hard glass, constructed by M. Chabaud, who also made the piezometers. A preliminary comparison of these instruments with the hydrogen thermometer showed only very small differences, for which allowance was made throughout. There is no room here to give the data for this calibration, as I had hoped to do; but I took into account the displacement of the zero mark by applying a special test after each series of measurements. The variations of this point were small.

However complicated the apparatus itself may appear, the measurements are made with facility, barring accidents, of course. The length of time consumed alone made them tedious. When a division mark has been brought into the field of the telescope, pressure is applied until the mark is tangent to the mercury meniscus, which here appears as a dark demarcation on a luminous background. The adjustment is easily made on compressing with the screw, and this, as in the preceding work, was used exclusively whenever the pressures exceeded a high value. Unfortunately, at high temperatures the action of the water eventually tarnishes the external surface of the graduated stem, and the meniscus appears blurred. Hence it is always necessary to stop after each series made at the higher temperatures, and take the apparatus apart in order to lightly repolish the stem. The operation is easily accomplished with the aid of a polishing wheel, mounted on a lathe. The present corrosion, however, is not to be compared to similar experiences with crown-glass mentioned above. It is for this reason that hard glass piezometers were selected.

Pressures are measured in the way described above. Obviously, account must be taken of the temperature of the mercury column, of the height of the mercury in the piezometer above its level in the barrel, and of the position of this level relatively to the manometer.

The measurement of small volumes is one of the difficulties of these researches. Without entering into any detail, I will simply state that allowance was made for the form of the meniscus, and of its position with reference to the division mark during calibration and during the subsequent measurements. Only in the case of gases are the errors here in question to be apprehended. Their volumes diminish with extreme rapidity, even when the reservoirs are made as large as is compatible with the dimensions of the barrel.

PART II.—DATA FOR GASES

The results summarized by the following tables were obtained in experiments of the kind just described.

After adding all corrections, the pressures were first expressed in atmospheres. To find the corresponding volumes, the piezometer may be supposed to be standardized at 0° C., seeing that volume ratios alone are in question. Account must be taken, however, when necessary, of the temperature differences occurring when the different parts of the piezometer (stem and reservoir) were calibrated. The corrections due to thermal expansion and to compression were afterwards applied in their turn. In my former researches I have given all the necessary data.

Having found the mass of the fluid in the manner stated above, and recalling that the calibration is supposed to be correct at zero, all the subsequent volumes are reduced to the value they would have if the given mass were that of unit of volume at 0° C. and 1 atmosphere. My tables, without exception, refer to this unit. Thereafter I constructed a series of curves corresponding to my data, in which pressures are the abscissas and the products, PV, of pressure and volume the ordinates. From these curves I selected a series of correlated values of PV, corresponding to groups of pressures differing in round numbers by 25, 50, or 100 atmospheres, and from them I deduced the values of V. For carbon dioxide and ethylene I batched the data in smaller pressure intervals because of the complex form of curve observed in the region of the critical point. Supplementary tables are here given.

The data for the pressure coefficient are taken directly from the curves. It sufficed to draw the lines of equal volume, which, under present circumstances, are straight lines passing through the origin, and then to read off the pressures at which these lines cut the successive isotherms. The curves were drawn either as a whole or by distributing them on sheets of millimetre cross-section paper, stretched on a large drawing-board more than 2 meters broad.

The gases studied are oxygen, hydrogen, nitrogen, air, carbon dioxide, and ethylene. The last were operated on by the method of sights only ; the other gases by both methods.

RESULTS OBTAINED BY THE FIRST METHOD

(*Method of Electric Contacts*)

I will begin with the results of the first method. The numbers relating to pressures below 500 atmospheres belong to the series of data obtained by the second method. The series found by the method of contacts does not begin until above 500 or 600 atmospheres. The other results are reproduced here so as to give completed series at zero.

TABLE 4

	OXYGEN			HYDROGEN			
	0° C.	0° C.	15.6° C.	0° C.	0° C.	15.4° C.	47.3° C.
P	PV	$V \times 10^6$	$V \times 10^6$	PV	$V \times 10^6$	$V \times 10^6$	$V \times 10^6$
Atm.							
1	1.0000	10^6	—	1.0000	10^6	—	—
100	.9265	9265	—	1.0690	10690	—	—
200	.9140	4570	—	1.1380	5690	—	—
300	.9625	3208	—	.1 2090	403000	—	—
400	1.0515	2629	—	1.2830	320700	—	—
500	1.1570	2314	—	1.3565	271300	—	—
600	1.2702	2117	2228	1.4322	2387	—	—
700	1.3867	1981	2075	1.5050	2150	2234	—
800	1.5040	1880	1959	1.5760	1970	2046	—
900	1.6200	1800	1871	1.6515	1835	1895	—
1000	1.7360	1736	1800	1.7250	1725	1778	1893
1100	1.8502	1682	1740	1.8007	1637	1685	1785
1200	1.9620	1635	1689	1.8690	1557.5	1604	1694.5
1300	2.0722	1594	1645	1.9383	1491	1533	1617.5
1400	2.1798	1557	1605	2.0048	1432	1472	1551
1500	2.2890	1526	1571	2.0700	1380	1418	1493
1600	2.3960	1497.5	1540	2.1352	1334.5	1370	1442
1700	2.5024	1472	1513.5	2.20065	1294.5	1326	1396

TABLE 4.—Continued

	OXYGEN			HYDROGEN			
	0° C.	0° C.	15.6° C.	0° C.	0° C.	15.4° C.	47.3° C.
P	PV	$V \times 10^6$	$V \times 10^6$	PV	$V \times 10^6$	$V \times 10^6$	$V \times 10^6$
Atm.							
1800	2.6073	1448.5	1488.5	2.2644	1258	1288	1354
1900	2.7113	1427	1465	2.3275	1225	1254.5	1316
2000	2.8160	1408	1444	2.3890	1194.5	1222.5	1280.5
2100	2.9190	1390	1424	2.44965	1166.5	1194	1249
2200	3.0217	1373.5	1406	2.5102	1141	1168.5	1220
2300	3.1234	1358	1390	2.5714	1118	1144.5	1194.5
2400	3.2244	1343.5	1374	2.6340	1097.5	1122.5	1170.5
2500	3.32375	1329.5	1360	2.6950	1078	1101	1148
2600	3.4229	1316.5	1346	2.7547	1059.5	1082.5	1126.5
2700	3.5208	1304	1332	2.8134	1042	1063	1107
2800	3.6176	1292	1319.5	2.8686	1024.5	1045	1088
2900	3.7120	1280	1307	—	—	1028	1071
3000	—	—	1296	—	—	1012.8	—

TABLE 5

	NITROGEN				AIR			
	0° C.	0° C.	16.0° C.	43.6° C.	0° C.	0° C.	15.7° C.	45.10° C.
P	PV	$V \times 10^6$	$V \times 10^6$	$V \times 10^6$	PV	$V \times 10^6$	$V \times 10^6$	$V \times 10^6$
Atm.								
1	1.0000	10⁶	—	—	1.0000	10⁶	—	—
100	.9910	9910	—	—	.9730	9730	—	—
200	1.0390	5195	—	—	1.0100	5030	—	—
300	1.1360	3786	—	—	1.0975	3658	—	—
400	1.2570	3142	—	—	1.2145	3036	—	—
500	1.3900	2780	—	—	1.3400	2680	—	—
600	1.5260	2543	—	—	1.4700	2450	—	—
700	1.6625	2375	—	—	1.6037	2291	2384	—
800	1.8016	2252	2331	—	1.7368	2171	2251.5	2387.5
900	1.9368	2152	2224	2354	1.8675	2075	2147	2271
1000	2.0700	2070	2134	2242	1.9990	1999	2061.5	2176.5
1100	2.20385	2003.5	2062	2162	2.1329	1939	1992	2097
1200	2.3352	1946	2000	2095	2.2596	1883	1933	2030
1300	2.46545	1896.5	1945	2035	2.3842	1834	1880	1970
1400	2.5942	1853	1897	1982	2.5081	1791.5	1834	1917
1500	2.72025	1813.5	1854	1933	2.6310	1754	1793.5	1871.5
1600	2.8456	1778.5	1818	1891.5	2.7528	1720.5	1757	1832.5
1700	2.9665	1745	1784	1853.5	2.87385	1690.5	1725	1796.5
1800	3.0861	1714.5	1752	1817.5	2.9916	1662	1695	1762.5
1900	3.20815	1688.5	1724.5	1787.5	3.1103	1637	1668	1733
2000	3.3270	1663.5	1699	1758.5	3.2260	1613	1643	1705
2100	3.4461	1641	1675	1731.5	3.34005	1590.5	1629	1678.5
2200	3.5640	1620	1653	1707	3.4540	1570	1598	1654
2300	3.6823	1601	1632	1683.5	3.56615	1550.5	1578	1632.5
2400	3.8004	1583.5	1613.5	1663.5	3.6804	1533.5	1559.5	1612
2500	3.9200	1568	1596	1644	3.79125	1516.5	1542	1593.5
2600	4.0378	1553	1579	1626	3.9000	1500	1525	1575.5
2700	4.1553	1539	1564	1608	4.00815	1484.5	1510	1557.5
2800	4.2700	1525	1549.5	1592	4.1146	1469.5	1495	1541
2900	4.3558	1502	1536	1577	4.2195	1455	1480.5	1525
3000	4.4970	1499	1522.5	1563	4.3230	1441	1466	1509.5

The following tables contain the data needed for the calculation of the pressure coefficients. They show the pressures at which the unit of mass occupies the volumes given in the first column at the stated temperatures:

TABLE 6.—PRESSURES AT CONSTANT VOLUME

	OXYGEN			HYDROGEN		
CONSTANT VOLUME	0° C.	15.6° C.	CONSTANT VOLUME	0° C.	15.4° C.	47.3° C.
$V \times 10^6$	Atm.	Atm.	$V \times 10^6$	Atm.	Atm.*	Atm.
2117	600	669	1725	1000	1055	1164
1880	800	888	1557.5	1200	1264	1390
1736	1000	1106	1380	1500	1579	1737
1635	1200	1325	1258	1800	1889	2071
1497.5	1600	1765	1194.5	2000	2099	2300
1408	2000	2188	1097.5	2400	2518	2746
1343.5	2400	2618	1024.5	2800	1925	—
1304	2700	2925				

	NITROGEN			AIR		
CONSTANT VOLUME	0° C.	16.0° C	43.6° C.	CONSTANT VOLUME	0° C.	15.7° C.	45.1° C.
$V \times 10^6$	Atm.	Atm.	Atm.	$V \times 10^6$	Atm.	Atm.	Atm.
2070	1000	1088	1239	2171	800	876	1007
1946	1200	1298	1474	1999	1000	1089	1250
1813.5	1500	1613	1812	1883	1200	1295	1474
1714.5	1800	1937	2168	1754	1500	1610	1828
1663.5	2000	2150	2401	1662	1800	1924	2166
1583.5	2400	2572	2858	1613	2000	2131	2394
1525	2800	2990	—	1533.5	2400	2552	2846

RESULTS OBTAINED BY THE SECOND METHOD

(*Method of Sights*)

The following tables of results obtained by the second method are arranged like the preceding, except that only the products PV are given for all the series:

TABLE 7.—OXYGEN

P	0° C.		15.65° C.		99.50° C.		199.5° C.	
	PV	$V \times 10^6$	PV	$V \times 10^6$	PV	$V \times 10^6$	PV	$V \times 10^6$
Atm.								
1	1.0000	10^6	—	—	—	—	—	—
100	.9265	9265	1.0045	10045	1.3750	13750	—	—
150	.9135	6090	.9920	6613	1.3820	9213	1.8000	12000
200	.9140	4570	.9945	4972	1.4000	7000	1.8190	9095
250	.9315	3726	1.0185	4054	1.4240	5696	1.8500	7400
300	.9625	3208	1.0420	3473	1.4530	4843	1.8850	6283
350	1 0040	2869	1.0800	3086	1.4900	4257	1.9220	5491
400	1.0515	2629	1.1250	2812	1.5320	3830	1.9610	4902
450	1.1025	2450	1.1750	2611	1.5760	3502	2.0040	4453
500	1.1560	2312	1.2270	2454	1.6220	3244	2.0500	4100
550	1.2120	2204	1.2815	2330	1.6690	3035	2.0950	3809
600	1.2690	2115	1.3370	2228	1.7200	2867	2.1420	3570
650	1.3275	2042	1.3940	2144	1.7725	2727	2.1910	3371
700	1.3855	1979	1.4515	2073	1.8270	2610	2.2415	3202
750	1.4440	1925	1.5080	2011	1.8810	2508	2 2920	3056
800	1.5030	1879	1.5660	1957	1.9340	2417	2.3430	2929
850	1.5615	1841	1 6240	1911	1.9875	2338	2.3950	2812
900	1.6200	1800	1.6820	1869	2.0415	2268	2.4465	2718
950	1.6780	1766	1.7400	1831	2.0960	2206	2.4980	2629
1000	1.7355	1735	1.7980	1798	2.1510	2151	—	—

TABLE 8.—HYDROGEN

P	0° C.		15.50° C.		99.25° C.		200.25° C.	
	PV	$V \times 10^6$	PV	$V \times 10^6$	PV	$V \times 10^6$	PV	$V \times 10^6$
Atm.								
1	1.0000	10^6	—	—	—	—	—	—
100	1.0690	10690	1.1290	11290	—	—	—	—
150	1.1030	7353	1.1630	7753	1.4770	9846	1.8480	12320
200	1.1380	5690	1.1980	5990	1.5135	7567	1.8840	9420
250	1.1730	4692	1.2350	4940	1.5500	6200	1.9200	7680
300	1.2090	4030	1.2685	4228	1.5860	5286	1.9560	6520
350	1.2460	3560	1.3050	3728	1.6225	4636	1.9930	5694
400	1.2830	3207	1.3410	3352	1.6590	4147	2.0300	5075
450	1.3200	2933	1.3780	3062	1.6950	3766	2.0670	4593
500	1.3565	2713	1.4150	2830	1.7310	3462	2.1050	4210
550	1.3935	2533	1.4520	2640	1.7675	3214	2.1400	3891
600	1.4315	2386	1.4890	2482	1.8040	3006	2.1762	3627
650	1.4685	2259	1.5260	2347	1.8400	2831	2.2120	3403
700	1.5045	2149	1.5620	2231	1.8760	2680	2.2480	3211
750	1.5400	2053	1.5985	2131	1.9130	2551	2.2840	3045
800	1.5775	1972	1.6340	2042	1.9490	2436	2.3200	2900
850	1.6140	1879	1.6690	1964	1.9860	2336	2.3560	2772
900	1.6490	1832	1.7060	1896	2.0210	2244	2.3915	2657
950	1.6850	1774	1.7410	1832	2 0660	2174	—	—
1000	1.7200	1720	1.7760	1776	2.0930	2093	—	—

TABLE 9.—NITROGEN

P	0° C. PV	0° C. V×10⁶	16.03° C. PV	16.03° C. V×10⁶	99.45° C. PV	99.45° C. V×10⁶	199.50° C. PV	199.50° C. V×10⁶
Atm.								
1	1.0000	10⁶	—	—	—	—	—	—
100	.9910	9910	1.0620	10620	—	—	—	—
150	1.0085	6723	1.0815	7210	1.4500	9666	1.8620	12410
200	1.0390	5195	1.1145	5572	1.4890	7445	1.9065	9532
250	1.0825	4330	1.1575	4630	1.5376	6150	1.9585	7834
300	1.1360	3786	1.2105	4035	1.5905	5301	2.0145	6715
350	1.1950	3414	1.2675	3621	1.6465	4703	2.0730	5923
400	1 2570	3142	1.3290	3322	1.7060	4265	2.1325	5331
450	1.3230	2940	1.3940	3098	1.7665	3924	2.1940	4875
500	1.3900	2780	1.4590	2918	1.8275	3655	2.2570	4514
550	1.4585	2652	1.5265	2775	1.8900	3436	2.3200	4218
600	1.5260	2543	1.5945	2657	1.9545	3258	2.3840	3973
650	1.5935	2452	1.6615	2556	2.0200	3108	2.4485	3613
700	1.6615	2374	1.7290	2470	2.0865	2980	2.5125	3589
750	1.7300	2307	1.7975	2397	2.1535	2871	2.5765	3435
800	1.7980	2247	1.8655	2332	2.2200	2775	2.6400	3300
850	1.8660	2195	1.9330	2274	2.2865	2690	2.7060	3184
900	1.9340	2149	2.0015	2224	2.3540	2616	2.7715	3079
950	2.0015	2107	2.0690	2178	2.4230	2550	2.8380	2987
1000	2.0685	2068	2.1360	2136	—	—	—	—

TABLE 10.—AIR

P	0 C. PV	0 C. V×10⁶	15.70° C. PV	15.70° C. V×10⁶	99.40° C. PV	99.40° C. V×10⁶	200.4° C. PV	200.4° C. V×10⁶
Atm.								
1	1.0000	10⁶	—	—	—	—	—	—
100	.9730	9730	1.0460	10460	1.4030	14030	—	—
150	.9840	6560	1.0580	7053	1.4310	9540	1.8430	12290
200	1.0100	5050	1.0855	5427	1.4670	7335	1.8860	9430
250	1.0490	4196	1.1260	4504	1.5110	6044	1.9340	7736
300	1.0975	3658	1.1740	3913	1.5585	5195	1.9865	6622
350	1.1540	3297	1.2250	3500	1.6085	4596	2.0410	5831
400	1.2145	3036	1.2835	3209	1.6625	4156	2 0960	5240
450	1.2765	2837	1.3460	2991	1.7200	3822	2.1530	4785
500	1.3400	2680	1.4110	2822	1.7815	3563	2.2110	4422
550	1.4040	2553	1.4740	2680	1.8440	3353	2.2700	4127
600	1.4700	2450	1.5375	2563	1.9060	3177	2.3300	3883
650	1.5365	2363	1.6015	2464	1.9670	3026	2.3900	3677
700	1.6020	2288	1.6670	2381	2.0300	2900	2.4515	3502
750	1.6690	2225	1.7340	2312	2.0930	2790	2.5130	3351
800	1.7345	2168	1.8000	2250	2.1555	2694	2.5750	3219
850	1.7990	2116	1.8655	2194	2.2180	2609	2.6370	3102
900	1.8640	2071	1.9300	2144	2.2830	2537	2.7000	3000
950	1.9280	2030	1.9960	2101	2.3490	2473	2.7640	2903
1000	1.9920	1992	2.0600	2060	2.4150	2415	2.8280	2828

I shall insert here also certain supplementary results obtained from these curves, which will be found useful in correcting the readings of gas manometers. To these I append the results of experiments made in 1864 at Fourvières with the same end in view:

TABLE 10 (2).—VOLUMES

(Same Unit of Mass as Heretofore)

P	OXYGEN AT 15.65° C. $V \times 10^6$	H YDROGEN AT 15.50° C. $V \times 10^6$	NITROGEN AT 16 05° C. $V \times 10^6$	AIR AT 15.70° C. $V \times 10^6$
125 *Atm.*	7976	9168	8560	8400
175 "	5663	6746	6255	6114
225 "	4456	5411	5044	4907
275 "	3735	4552	4298	4178
325 "	3261	3959	3790	3689
375 "	2939	3528	3460	3343
425 "	2706	3199	3205	3094
475 "	2528	2940	3003	2900

TABLE 11.—EXPERIMENTS AT FOURVIÈRES

(Values of PV at 16°)

PRESSURES IN METERS	NITROGEN	AIR
.76	1.0000	1.0000
20	.9930	.9901
25	.9919	.9876
30	.9908	.9855
35	.9899	.9832
40	.9896	.9824
45	.9895	.9815
50	.9897	.9808
55	.9902	.9804
60	.9908	.9803
65	.9913	.9807

TABLE 12.—DATA FOR THE COMPUTATION OF THE COEFFICIENTS OF PRESSURE

(*Pressure at Constant Volume*)

CONSTANT VOLUME	OXYGEN				CONSTANT VOLUME	HYDROGEN			
	0° C.	16.65°C.	99.50°C.	199.50° C.		0° C.	15.5°C.	99.25°C.	200.25° C.
$V \times 10^6$	Atm.	Atm.	Atm.	Atm.	$V \times 10^6$	Atm.	Atm.	Atm.	Atm.
9265	100	108	149	196	10690	100	106	137	174
6090	150	163	233	312	7353	150	159	207	262
4570	200	219	322	437	5690	200	212	276	351
3726	250	276	415	566	4692	250	265	345	439
3208	300	332	508	698	4030	300	318	414	528
2869	350	388	598	827	3560	350	370	482	614
2629	400	446	691	—	3207	400	423	551	700
2450	450	502	781	—	2933	450	476	620	788
2312	500	558	868	—	2713	500	530	688	874
2204	550	624	953	—	2533	550	582	756	—
					2386	600	635	824	—
					2259	650	687	891	—
					2149	700	741	960	—

CONSTANT VOLUME	NITROGEN				CONSTANT VOLUME	AIR			
	0° C.	16.03°C.	99.45°C.	199.5° C.		0° C.	15.70°C.	99.40°C.	200.40° C.
$V \times 10^6$	Atm.	Atm.	Atm.	Atm.	$V \times 10^6$	Atm.	Atm.	Atm.	Atm.
9910	100	107	146	192	9730	100	107	146	193
6723	150	162	225	299	6560	150	162	227	303
5195	200	217	307	414	5050	200	217	310	420
4330	250	273	392	530	4196	250	373	395	538
3786	300	328	474	644	3658	300	329	479	655
3414	350	383	556	758	3297	350	383	564	770
3142	400	439	637	869	3036	400	439	646	881
2940	450	494	718	—	2837	450	495	728	993
2780	500	548	797		2680	500	550	807	—
2652	550	602	875		2553	550	603	887	—
2543	600	656	957		2450	600	658	970	—

The following tables relative to carbon dioxide and ethylene contain the values of the products PV only :

.84

THE LAWS OF GASES

TABLE 13.—VALUES OF PV FOR CARBON DIOXIDE

P	0° C.	10° C.	20° C.	30° C.	40° C.	50° C.	60° C.
Atm.							
1	1.0000	—	—	—	—	—	—
50	.1050	.1145	.6800	.7750	.8500	.9200	.9840
75	.1530	.1630	.1800	.2190	.6200	.7470	.8410
100	.2020	.2130	.2285	.2550	.3090	.4910	.6610
125	.2490	.2620	.2785	.3000	.3350	.3950	.5100
150	.2950	.3090	.3260	.3460	.3770	.4190	.4850
175	.3405	.3550	.3725	.3930	.4215	.4570	.5055
200	.3850	.4010	.4190	.4400	.4675	.5000	.5425
225	.4305	.4455	.4655	.4875	.5130	.5425	.5825
250	.4740	.4900	.5100	.5335	.5580	.5865	.6250
275	.5170	.5340	.5545	.5775	.6040	.6330	.6675
300	.5595	.5775	.5985	.6225	.6485	.6765	.7100
350	.6445	.6640	.6850	.7090	.7365	.7650	.7980
400	.7280	.7475	.7710	.7950	.8230	.8515	.8840
450	.8090	.8310	.8550	.8800	.9075	.9365	.9690
500	.8905	.9130	.9380	.9630	.9900	1.0210	1.0540
550	.9700	.9935	1.0200	1.0465	1.0740	1.1035	1.1370
600	1.0495	1.0730	1.0995	1.1275	1.1570	1.1865	1.2190
650	1.1275	1.1530	1.1800	1.2075	1.2375	1.2680	1.3010
700	1.2055	1.2320	1.2590	1.2890	1.3190	1.3500	1.3825
750	1.2815	1.3105	1.3395	1.3700	1.4000	1.4315	1.4640
800	1.3580	1.3870	1.4170	1.4475	1.4790	1.5105	1.5435
850	1.4340	1.4625	1.4935	1.5245	1.5570	1.5885	1.6225
900	1.5090	1.5385	1 5685	1.6000	1.6325	1.6650	1.6995
950	1.5830	1.6115	1.6430	1.6740	1.7065	1.7395	1.7745
1000	1.6560	1.6850	1.7160	1.7480	1.7800	1.8140	1.8475

P	70° C.	80° C.	90° C.	100° C.	137° C.	198° C.	258° C.
Atm.							
1	—	—	—	—	—	—	—
50	1.0430	1.0960	1.1530	1 2065	1.3800	—	—
75	.9180	.9380	1.0515	1.1180	1.3185	1.6150	1.8670
100	.7770	.8725	.9535	1.0300	1.2590	1 5820	1.8470
125	.6430	.7590	.8580	.9470	1.2050	1.5530	1.8310
150	.5750	.6805	.7815	.8780	1.1585	1.5295	1.8180
175	.5730	.6515	.7410	.8320	1.1230	1.5100	1.8095
200	.5955	.6600	.7315	.8145	1.0960	1.4960	1.8040
225	.6285	.6815	.7460	.8175	1.0835	1.4890	1.8035
250	.6670	.7135	.7690	.8355	1.0810	1.4870	1.8060
275	.7070	.7515	.8015	.8600	1.0885	1.4875	1.8115
300	.7485	.7900	.8375	.8900	1.1080	1.4935	1.8200
350	.8325	.8725	.9135	.9615	1.1565	1.5210	1.8465
400	.9180	.9560	.9660	1.0385	1.2175	1.5630	1.8830
450	1.0035	1.0400	1.0775	1.1190	1.2880	1.6160	1.9280
500	1.0880	1.1240	1.1610	1.2005	1.3620	1.6775	—
550	1.1720	1.2085	1.2430	1.2830	1.4400	1.7450	—
600	1.2540	1.2900	1.3265	1.3655	1.5180	1.8120	—
650	1.3360	1.3725	1.4085	1.4475	1.5960	1.8835	—
700	1.4170	1.4535	1.4900	1.5285	1.6760	1.9560	—
750	1.4985	1.5335	1.5705	1.6100	1.7565	2.0330	—
800	1.5780	1.6140	1.6505	1.6890	1.8355	2 1080	—
850	1.6575	1.6925	1.7285	1.7680	1 9150	2 1860	—
900	1.7345	1.7710	1.8075	1.8460	1.9940	2.2600	—
950	1.8100	1 8470	1.8845	1.9230	2.0720	2.3350	—
1000	1.8840	1.9210	1.9590	1.9990	—	—	—

TABLE 14.—CARBON DIOXIDE

(Supplementary Table for Values of PV)

P	0°	10°	20°	30°	32°	35°	40°	50°	60°	70°	80°	90°	100°
Atm.													
31	.7380	—	—	—	—	—	—	—	—	—	—	—	—
33	.7120	.7860	—	—	—	—	—	—	—	—	—	—	—
34	.6990	.7750	—	—	—	—	—	—	—	—	—	—	—
35	.0750	.7640	.8350	—	—	—	—	—	—	—	—	—	—
37	.0790	.7420	.8170	.8820	—	—	—	—	—	—	—	—	—
40	—	.7060	.7895	.8590	.8750	.8920	.9235	—	—	—	—	—	—
44	—	.6530	.7490	—	—	—	—	—	—	—	—	—	—
45	—	.1050	.7380	.8190	.8350	.8555	.8880	.9520	1.0110	1.0660	—	—	—
48	—	—	.7060	.7930	—	—	.8670	.9330	.9950	1.0520	—	—	—
50	1.1050	.1145	.6800	.7750	.7920	.8155	.8525	.9210	.9840	1.0430	1.0980	1.1535	1.2070
53	—	—	.6370	.7460	—	—	.8300	.9020	.9680	1.0280	1.0850	1.1420	1.1960
55	—	—	.6050	.7260	.7455	.7720	.8135	.8890	.9570	1.0185	1.0760	1.1340	1.1785
56	—	—	.5850	—	—	—	—	—	—	—	—	—	—
57	—	—	.1480	—	—	—	—	—	—	—	—	—	—
60	—	—	.1520	.6680	.6935	.7245	.7720	.8555	.9285	.9940	1.0540	1.1130	1.1710
65	—	—	—	.5950	.6290	.6690	.7260	.8200	.8990	.9690	1.0325	1.0930	1.1530
68	—	—	—	.5350	.5780	.6310	.6950	.7970	.8810	.9530	1.0190	1.0810	1.1420
70	—	—	—	.4700	.5400	.6020	.6730	.7820	.8685	.9430	1.0100	1.0730	1.1350
71	—	—	—	.2300	—	—	—	—	—	—	—	—	—
72	—	—	—	.2230	.4910	—	—	—	—	—	—	—	—
73	—	—	—	—	.4600	—	—	—	—	—	—	—	—
74	—	—	—	.2190	.4050	.5310	—	—	—	—	—	—	—
74.5	—	—	—	—	.3400	—	—	—	—	—	—	—	—
75	—	—	—	.2190	.2680	.5100	.6130	.7410	.8360	.9170	.0880	1.0535	1.1180
76	—	—	—	—	.2495	.4850	—	—	—	—	—	—	—
78	—	—	—	.2205	.2410	.4200	—	—	—	—	—	—	—
80	—	—	—	.2225	—	.3180	.5400	.7000	.8030	.8900	.9660	1.0335	1.1005
82	—	—	—	—	—	.2810	.5030	—	—	—	—	—	—
85	—	—	—	—	—	.2670	.4350	.6510	.7690	.8630	.9425	1.0135	1.0835
90	—	—	—	—	—	.2650	.3410	.5990	.7340	.8350	.9190	.9935	1.0655
95	—	—	—	—	—	—	.3140	.5460	.6980	.8060	.8960	.9735	1.0480
100	—	—	—	—	—	—	.3090	.4910	.6610	.7770	.8720	.9540	1.0305
110	—	—	—	—	—	—	.3130	.4170	.5880	.7210	.8240	.9140	.9070

[For Table 15, see p. 87.]

TABLE 16.—SUPPLEMENTARY VALUES OF PV FOR ETHYLENE

P	0°	5°	7.5°	10°	20°	30°	40°	50°	60°	70°	80°	90°	100°
Atm.													
36	.6340	—	—	—	—	—	—	—	—	—	—	—	—
37	.6165	—	—	—	—	—	—	—	—	—	—	—	—
38	.5955	.6490	.6735	—	—	—	—	—	—	—	—	—	—
39	—	—	—	.6820	—	—	—	—	—	—	—	—	—
40	.5330	.6155	.6425	.6685	—	—	—	—	—	—	—	—	—
41	.1610	—	—	—	—	—	—	—	—	—	—	—	—
42	.1570	.5730	.6085	.6370	.7320	—	—	—	—	—	—	—	—
43	.1580	.5470	—	—	—	—	—	—	—	—	—	—	—
44	.1600	.5150	.5675	.6030	—	—	—	—	—	—	—	—	—
45	—	.4770	—	—	.6980	—	—	—	—	—	—	—	—
46	.1645	.1890	.5100	.5620	.6840	—	—	—	—	—	—	—	—
47	—	.1850	.4670	—	—	—	—	—	—	—	—	—	—
48	.1095	.1855	.3900	.5075	—	—	.8300	—	—	—	—	—	—
49	—	.1875	.2150	.4700	—	—	—	—	—	—	—	—	—
50	.1755	.1900	.2075	.4200	.6290	.7310	.8140	.8865	—	—	—	—	—
51	—	—	—	.2900	—	—	—	—	—	—	—	—	—
52	.1810	.1945	.2060	.2400	.5975	—	—	—	—	—	—	—	—
53	—	—	—	—	—	—	—	—	—	—	—	—	—
54	—	—	.2090	.2290	.5610	.6905	.7810	.8595	.9290	—	—	—	—
56	—	.2050	.2125	.2270	.5235	—	—	—	—	.9850	—	—	—
58	—	—	.5180	.2285	.4805	—	—	—	—	—	—	—	—
60	.2025	.2145	—	.2315	.4300	.6195	.7285	.8170	.8925	.9630	1.0285	1.0920	1.1530
65	—	—	—	—	.3310	.5500	.6805	—	—	—	—	—	—
70	—	—	—	—	.3110	.4830	.6310	.7430	.8315	.9090	.9795	—	—
75	.2425	.2535	—	—	.2655	.3110	.4300	.6805	—	.8815	.9550	1.0260	1.0940
80	.2565	—	—	.2785	.3185	.3990	.5390	.6660	.7670	.8555	.9310	1.0050	1.0755
90	—	—	—	.3370	.3915	.4875	.6060	.7090	.8035	.8840	—	—	—
100	.3100	—	—	.3305	.3500	.4030	.4710	.5665	.6680	.7620	.8465	.9265	1.0050

TABLE 15.—VALUES OF PV FOR ETHYLENE

P Atm.	0°	10°	20°	30°	40°	50°	60°	70°	80°	90°	100°	137.5°	198.5°
1	1.0000	—	—	—	—	—	—	—	—	—	—	—	—
50	.1755	.4200	.6290	.7310	.8140	.8875	.9535	1.0175	1.0770	1.1350	1.1920	1.3736	1.6520
75	.2425	.2655	.3110	.4300	.5805	.7045	.8000	.8815	.9550	1.0260	1.0940	1.3056	1.6140
100	.3100	.3305	.3600	.4030	.4705	.5665	.6680	.7620	.8465	.9265	1.0050	1.2466	1.5800
125	.3750	.3950	.4225	.4550	.4990	.5550	.6245	.7090	.7920	.8690	.9450	1.2030	1.5540
150	.4405	.4590	.4850	.5150	.5505	.5945	.6490	.7090	.7760	.8490	.9240	1.1780	1.5400
175	.5040	.5230	.5475	.5770	.6090	.6480	.6920	.7435	.7980	.8575	.9245	1.1670	1.5350
200	.5650	.5850	.6095	.6380	.6690	.7030	.7440	.7895	.8380	.8900	.9460	1.1670	1.5350
225	.6270	.6475	.6720	.6995	.7285	.7615	.7985	.8401	.8855	.9345	.9860	1.1740	1.5368
250	.6870	.7080	.7325	.7590	.7880	.8200	.8560	.8955	.9370	.9830	1.0315	1.1956	1.5490
300	.8055	.8270	.8520	.8780	.9075	.9390	.9720	1.0085	1.0475	1.0865	1.1330	1.2284	1.5690
350	.9229	.9440	.9690	.9955	1.0250	1.0555	1.0875	1.1205	1.1580	1.1990	1.2420	1.3100	1.6270
400	1.0365	1.0585	1.0840	1.1115	1.1405	1.1705	1.2020	1.2350	1.2725	1.3135	1.3560	1.4060	1.7010
450	1.1465	1.1705	1.1975	1.2255	1.2550	1.2855	1.3175	1.3505	1.3865	1.4255	1.4660	1.5104	1.7900
500	1.2555	1.2800	1.3075	1.3370	1.3670	1.3985	1.4310	1.4645	1.5000	1.5360	1.5775	1.6150	1.8858
550	1.3640	1.3895	1.4165	1.4465	1.4770	1.5090	1.5420	1.5765	1.6115	1.6470	1.6855	1.7212	1.9846
600	1.4725	1.4985	1.5250	1.5555	1.5865	1.6180	1.6520	1.6865	1.7215	1.7570	1.7950	1.8290	2.0868
650	1.5785	1.6035	1.6325	1.6630	1.6930	1.7265	1.7610	1.7950	1.8305	1.8665	1.9035	1.9376	2.1910
700	1.6835	1.7090	1.7375	1.7680	1.7995	1.8330	1.8670	1.9015	1.9365	1.9735	2.0115	2.0450	2.2950
750	1.7865	1.8130	1.8420	1.8725	1.9050	1.9385	1.9720	2.0080	2.0420	2.0800	2.1190	2.1526	2.3990
800	1.8880	1.9165	1.9460	1.9770	2.0100	2.0435	2.0775	2.1130	2.1495	2.1865	2.2245	2.2604	2.5030
850	1.9900	2.0190	2.0495	2.0815	2.1140	2.1475	2.1820	2.2185	2.2555	2.2925	2.3300	2.3684	2.6060
900	2.0905	2.1215	2.1530	2.1850	2.2175	2.2505	2.2865	2.3225	2.3595	2.3970	2.4345	2.4762	2.7104
950	2.1900	2.2215	2.2535	2.2865	2.3200	2.3545	2.3900	2.4265	2.4635	2.5005	2.5390	2.5848	2.8140
1000	2.2890	2.3205	2.3535	2.3870	2.4215	2.4565	2.4925	2.5290	2.5660	2.6035	2.6425	2.6916	—

TABLE 17.—CARBON DIOXIDE. DATA FOR THE COMPUTATION OF PRESSURE COEFFICIENTS

(Pressures at Constant Volume)

CONSTANT VOLUME $V \times 10^6$	0°	10°	20°	30°	40°	50°	60°	70°	80°	90°	100°	137°	198°	268°
	Atm.	Atm.	Atm.	Atm.	Atm.	Atm.	Atm.	Atm.	Atm.	Atm.	Atm.	Atm.	Atm.	Atm.
23850	31.0	33.0	35.0	37.0	39.0	40.9	42.8	44.7	46.6	48.5	50.5	57.0	68.0	78.5
16360	34.4	41.8	45.1	48.3	51.4	54.5	57.6	60.6	63.5	66.5	69.5	80.0	97.0	112.0
13000	34.4	44.4	51.1	55.5	59.7	63.8	67.8	71.8	75.7	79.6	83.6	97.5	120.0	140.0
10000	34.4	44.4	56.4	62.8	68.6	74.5	80.2	85.8	91.3	96.7	102.3	121.5	153.5	181.0
7690	34.4	44.4	56.4	68.3	76.6	84.8	92.8	100.6	108.2	116.0	123.8	151.0	195.0	234.5
5780	34.4	44.4	56.4	70.7	83.1	94.7	106.2	117.5	128.8	140.2	151.3	191.0	257.0	316.0
4280	34.4	44.4	56.4	70.7	87.8	104.8	121.9	138.9	156.3	173.5	191.1	252.5	356.0	449.5
3160	34.4	44.4	56.4	71.5	98.0	125.3	153.8	183.2	211.5	240.5	271.0	376.0	554.5	—
2300	—	—	64.4	109.0	155.0	201.0	250.5	298.5	340.0	394.5	443.5	619.0	909.0	—
2000	122.5	209.0	300.0	384.0	470.5	560.0	651.0	745.0	832.5	918.0	998.0	—	—	—
1870	307.5	404.0	520.0	627.5	750.5	856.5	953.5	—	—	—	—	—	—	—

TABLE 18.—ETHYLENE. DATA FOR THE COMPUTATION OF PRESSURE COEFFICIENTS

(Pressures at Constant Volume)

CONSTANT VOLUME $V \times 10^6$	0°	10°	20°	30°	40°	50°	60°	70°	80°	90°	100°	137.5°	198.5°
	Atm.	Atm.	Atm.	Atm.	Atm.	Atm.	Atm.	Atm.	Atm.	Atm.	Atm.	Atm.	Atm.
16666	37.0	40.0	43.2	46.2	49.2	52.2	55.2	58.3	61.3	64.3	67.3	78.0	95.4
11500	40.6	47.2	52.5	57.7	62.6	67.6	72.5	77.3	82.3	87.0	91.8	109.0	137.0
8333	40.6	50.2	57.9	65.4	72.6	80.0	87.0	94.2	101.2	108.2	115.6	141.7	183.7
6428	40.6	50.8	61.3	71.8	81.9	92.5	102.6	113.1	123.5	133.7	144.3	182.5	244.5
5000	40.6	51.2	65.4	80.0	95.0	110.0	125.0	140.8	153.5	170.7	186.0	243.3	334.7
4166	40.6	54.6	74.4	94.8	116.0	136.4	150.4	180.6	202.0	223.3	246.4	325.3	454.5
3500	49.0	77.2	108.5	140.0	171.0	203.0	234.5	266.0	298.0	329.5	364.2	478.5	663.0
3000	124.5	169.2	219.5	265.0	311.7	358.0	406.0	451.0	499.0	546.0	595.0	760.5	—
2857	185.5	236.5	288.4	341.0	395.0	449.0	499.0	551.0	606.5	660.5	716.0	—	—

EXAMINATION OF THE RESULTS

General Laws

An inspection of the above results will lead to inferences similar in a general way to those which I adduced in my memoir of 1881. I shall now, moreover, be able to examine a

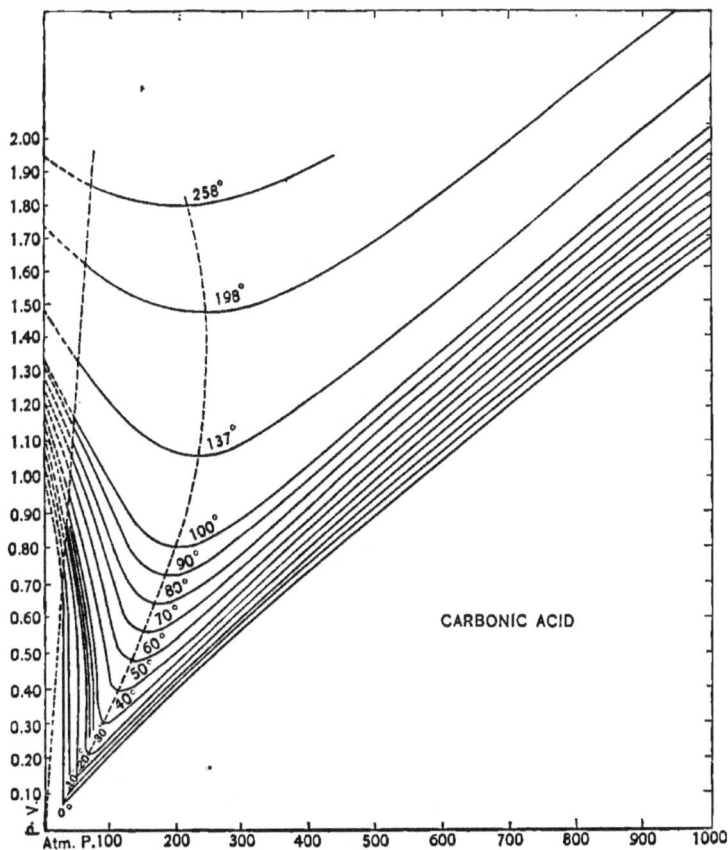

FIG. 9.

great number of moot points, relative to which no decision could certainly have been made by aid of investigations no more extensive than those heretofore available.

It did not seem necessary to give all the curves for the divers gases, seeing that the principal types were drawn in full in my first research. The group of isotherms for carbon dioxide graphically represented within the limits actually reached suffice for exhibiting the results in their general features. The first diagram (Fig. 9) shows the isotherms for this gas, when the pressures are laid off as abscissas and the products PV as ordinates. The second diagram (Fig. 10) shows the region in

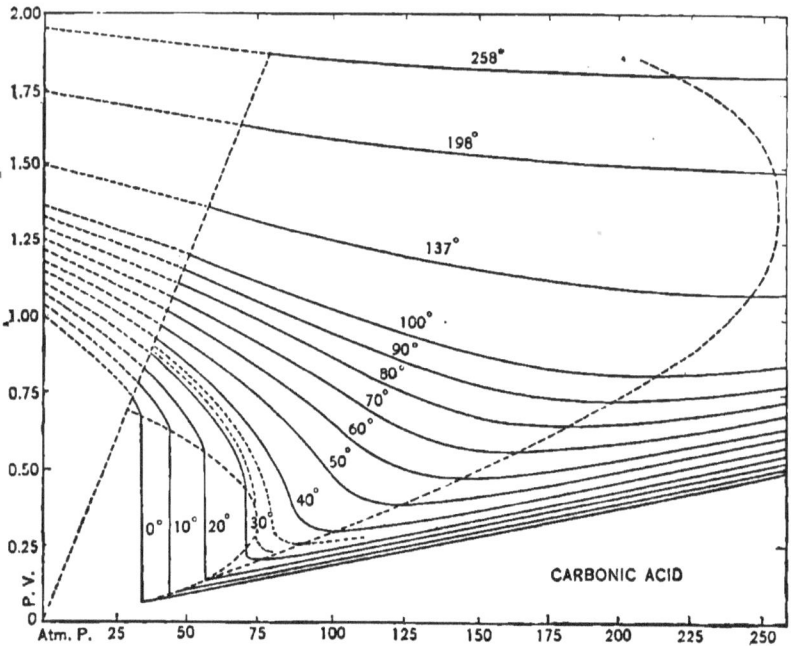

FIG. 10.

the neighborhood of the critical point with more detail. The part of the isotherms represented by dotted lines has been added merely to round off the figures. They have not been made the subject of measurement in the present paper, excepting the lines for 32° and 35°, which are dotted throughout.

At temperatures lower than the critical point, a part of the isotherms is in the form of straight vertical lines, corresponding to liquefaction. These did not occur in my first group of curves, which began at 35°.

The ordinates of the extremities of each straight part show

the volume in the liquid state and the corresponding volume of the saturated vapor, and hence, also, the densities of the two states, respectively. The locus of these points is Andrews's curve of liquefaction, and has been traced in the second diagram (Fig. 10). Another dotted curve, recalling, as does the preceding, the form of a parabola between the limits of construction, is the locus of the points of minimum ordinates, PV. The abscissas of this curve pass through a maximum, as was to be anticipated from the facts which I have already indicated elsewhere—*i.e.*, for gases in a region remote from their critical points, like methane, air, nitrogen, the abscissa of the minimum ordinate shows a retrograde march for continually increasing temperatures, an inversion of what took place in the case of carbon dioxide within the field to which I was then restricted. The following table gives the maximum vapor tension, the pressure corresponding to the minimum ordinates at the different temperatures, as well as the value of PV at that ordinate for the gases oxygen, carbon dioxide, and ethylene :

TABLE 19

	VALUES OF PV						VAPOR TENSION	
	CARBON DIOXIDE		ETHYLENE		OXYGEN		CARB. DIOX.	ETHYLENE
T	P	PV	P	PV	P	PV	P	P
$C.$	*Atm.*		*Atm.*		*Atm.*		*Atm.*	*Atm.*
0°	34.5	.0740	42	.1570	175	.9120	34.4	40.6
5	—	—	47.5	.1850	—	—	—	45.5
7.5	—	—	51.5	.2055	—	—	—	48.1
10	44.5	.1035	55.7	.2270	—	—	44.4	51.1
15	—	—	—	—	165	.9910	—	—
20	56.8	.1475	72	.3095	—	—	56.4	—
30	76	.2185	87	.3900	—	—	70.7	—
40	101	.3083	101	.4700	—	—		
50	125	.3465	114	.5485	—	—		
60	143	.4830	125	.6245	—	—		
70	162	.5690	135	.7030	—	—		
80	179	.6500	145	.9750	—	—		
90	196	.7310	153	.8490	—	—		
100	210	.8140	161	.9220	100	1.3750		
137	245	1.0850	185	1.1660	—	—		
198	255	1.4920	188	1.5340	—	—		
258	218	1.8100	—	—	—	—		

The temperature 10° C., corresponding to the maximum vapor pressure 51 atmospheres, appears from the form of the isotherms for ethylene to be extremely near the critical temperature. These results are thus approximately identical with the data obtained by Mr. J. Dewar. The method pursued does not admit of a direct determination of the point in question, and critical data can be deduced from the table only by calculation based on the intrinsic equation of the gas.

Quite recently M. Mitkowski,* in an interesting paper on the compressibility of air, has prolonged the locus of minimum ordinates as far as 145° C., and has shown that this gas has a maximum abscissa at about −75° C. and 124 atmospheres.

The form of the isotherms beyond the region of minimum ordinates is one of the questions which I particularly wished to investigate. These curves, beginning with a distance from the ordinates in question, which is smaller in proportion as the temperature is lower, seemed to me to undergo a transformation into lines sensibly straight. I was aware of the existence of slight curvature; but this was so little marked as to suggest that further prolongation of the family of curves would bring out a fascicle of parallel straight lines more and more clearly. In fact, this hypothesis proved to be specially attractive, inasmuch as the angular coefficient of these fascicles severally gave the limit of volume for an infinite pressure. Hence they lead to a very simple interpretation of the *covolume*, thus found directly and with precision.

Unfortunately, this hypothesis of mine does not seem to be verified—at least. within the limits of temperature and pressure of the present research. The isotherms all present a concavity towards the abscissa, slight, it is true, but nevertheless beyond question. Concavity is expressible as a diminution of the angular coefficient of the tangent, and I have summarized the values of the coefficient $\dfrac{P'V'-PV}{P'-P} = \epsilon$ in the following table. This gives ϵ between the pressure limits given in the first column at the different temperatures indicated in the first row.

* Académie des Sciences de Cracovie, 1891.

TABLE 20.—VALUES OF $\varepsilon = \dfrac{P'V' - PV}{P' - P}$

		HYDROGEN $\varepsilon \times 10^6$ AT		NITROGEN $\varepsilon \times 10^6$ AT		AIR $\varepsilon \times 10^6$ AT		OXYGEN $\varepsilon \times 10^6$ AT
		0° C.	47.3°	0° C.	43.6°	0° C.	45.1°	0° C.
Atm.	*Atm.*							
From 500 to 1000..		732	—	—	—	—	—	1158
" 1000 " 1500..		690	693	1300	1316	1264	1261	1106
" 1500 " 2000..		638	643	1213	1233	1190	1206	1054
" 2000 " 2500..		612	618	1186	1176	1301	1147	1015
" 2500 " 3000..		579	588	1154	1168	1063	1090	971

		HYDROGEN $\varepsilon \times 10^6$ AT			ETHYLENE $\varepsilon \times 10^6$ AT		CARBON DIOXIDE $\varepsilon \times 10^6$ AT	
		0° C.	99.25°	200.5°	0° C.	100 0°	0° C.	100.0°
Atm.	*Atm.*							
From 200 to 400.....		725	727	730	2357	—	1715	—
" 400 " 600.....		742	725	732	2180	2195	1607	1635
" 600 " 800.....		730	725	719	2080	2157	1542	1617
" 800 " 1000.....		712	720	—	2002	2090	1490	1550

Clearly the isotherms present slight curvature, as pointed out above. Between the same pressure limits the angular coefficient increases by a small amount with temperature. This increment corresponds to widening of the fascicle, which is distinctly seen for the case of carbon dioxide in the region of lower temperatures. As temperature increases the curves gradually cease to spread apart. In the permanent gases—like hydrogen, nitrogen, and air—the variation with temperature is scarcely perceptible.

A comparison of the decrements of these coefficients between the same pressure limits but at different temperatures shows no variation clearly enough indicated to be specified like the preceding; those groups of observations which extend over a sufficient interval of temperature are restricted to a pressure interval of 1000 atmospheres. Whether under sufficiently high pressures and at high enough temperatures the angular coefficient will reach a limiting value cannot, therefore, be foreseen. In all cases the smallest values of these coefficients are superior limits of the smallest volumes possible. It might be interesting to compare these values with those computed by aid of the intrinsic equations. To take the simplest form of equation, that of Van der Waals, the third part of the critical volume is evidently a limit of this reduction.

Carbon dioxide, for instance, has a critical volume of .004224, the third of which, .001408, is markedly less than the smallest angular coefficient, .00149.

COEFFICIENTS OF EXPANSION AT CONSTANT PRESSURE

$$\left(\frac{1}{v}\frac{dv}{dt}\right)$$

The laws of expansion are particularly involved in the neighborhood of the critical point. For temperatures below the critical point the coefficients can obviously not be computed for pressures intermediate between the tension maxima corresponding to the given limits of temperature; for the change of volume here originates not merely in thermal expansion between these states, but is due also to a change of state. In other words, the coefficients are infinite between these limits, and in the following table crosses are put in the place of the two coefficients which, for the reason given, are without meaning.

For pressures equal to the maximum vapor tensions at one of the limits of the temperature interval, the coefficient refers to the gaseous state for the case of the lower limit, and to the liquid state for the case of the upper limit. Hence, for a given temperature and at the corresponding maximum vapor tension, there are two coefficients, belonging, respectively, to the two states of aggregation. One refers to incipient saturation, the other to an absence of vapor. It would be extremely interesting to compare the values of these two coefficients at different temperatures. Such an inquiry, however, would require special investigations, and on the whole present serious difficulties. Since the variation of the coefficient with temperature is very rapid under these conditions, it would be necessary to greatly restrict the temperature interval on approaching saturation.

Two tables follow relative to carbon dioxide and ethylene respectively, in which the mean coefficients of expansion are given for the pressures inserted in the first vertical column. The temperature intervals for which the coefficients apply are shown in the first horizontal row. To avoid misapprehension, it is to be observed that the coefficients are reduced or referred to unit of volume by successively dividing by the volume corresponding to the lower temperature limit, and not by the initial volume at zero centigrade.

TABLE 21.—VALUES OF $\dfrac{1}{v}\dfrac{\Delta v}{\Delta t}=a$ FOR CARBON DIOXIDE

P	0°-10°	10°-20°	20°-30°	30°-40°	40°-50°	50°-60°	60°-70°	70°-80°	80°-90°	90°-100°	100°-137°	137°-198°	198°-258°
Atm.	$a\times10^5$	$a\times10^5$	$a\times10^5$	$a\times10^5$	$a\times10^5$	$a\times10^5$	$a\times10^5$	$a\times10^5$	$a\times10^5$	$a\times10^5$	$a\times10^5$	$a\times10^5$	$a\times10^5$
50	905	↑	1394*	1097*	823	695	600	508	520	464	—	—	—
60	800	1259	↑	1557*	1081	853	705	604	560	521	465	—	—
75	654	1043	2166	[(18310)]	2048	1258	916	762	643	595	—	369	260
80	644	951	1710	14270	2963	1471	1083	854	699	648	—	—	—
85	637	872	1425	9079	4965	1813	1222	921	753	696	—	—	—
90	632	827	1265	4450	[(75666)]	2254	1376	1006	811	724	—	—	—
95	626	786	1162	2900	[(7388)]	2784	1547	1116	865	765	601	420	279
100	544	728	1159	2128	[(6899)]	3462	1755	1229	928	802	—	—	—
110	544	709	923	1510	3142	[(4100)]	[(2911)]	1428	1092	908	—	—	—
125	522	630	772	1666	1791	1575	[(2608)]	1604	1304	1037	—	—	—
150	474	550	613	922	1114	1061	1855	[(1835)]	[(1484)]	[(1247)]	864	525	313
175	423	493	550	728	842	850	1335	1370	[(1374)]	1228	(945)	565	330
200	416	449	501	625	695	656	977	1083	1083	1134	934	798	343
250	337	408	461	459	511	495	672	697	778	[865]	794	(616)	358
300	322	364	401	418	432	382	542	554	601	627	[662]	570	(364)
400	268	314	311	352	346	323	384	414	419	426	[466]	465	341
500	253	274	266	280	313	274	322	330	329	349	364	386	—
600	224	247	255	261	255	241	287	287	283	293	302	317	—
700	220	219	238	232	235	219	249	258	251	258	261	278	—
800	214	216	215	218	223	210	224	228	226	233	234	243	—
900	197	195	200	204	199	210	206	210	206	213	216	223	—
1000	175	184	180	163	191	184	198	197	198	204	234	—	—

TABLE 22.—VALUES OF $\dfrac{1}{v}\dfrac{\Delta v}{\Delta t} = \alpha$ FOR ETHYLENE

P Atm.	0°-10°	10°-20°	20°-30°	30°-40°	40°-50°	50°-60°	60°-70°	70°-80°	80°-90°	90°-100°	100°-137.5°	137.5°-198.5°
	$\alpha\times10^5$	$\alpha\times10^5$	$\alpha\times10^5$	$\alpha\times10^5$	$\alpha\times10^5$	$\alpha\times10^5$	$\alpha\times10^5$	$\alpha\times10^5$	$\alpha\times10^5$	$\alpha\times10^5$	$\alpha\times10^5$	$\alpha\times10^6$
50	—	4976	1622	1135	902	744	671	585	539	502	406	332
75	948	1714	([3826])	(3500)	(2136)	1355	1019	834	743	663	516	387
100	661	893	1195	1427	[2040]	(1790)	(1407)	1109	945	847	641	438
125	533	696	769	967	1122	1252	[1353]	(1171)	(972)	875	728	478
150	420	566	619	689	799	917	925	[945]	941	(883)	(733)	504
175	377	468	539	555	640	679	744	746	746	[781]	699	517
200	354	419	467	486	508	583	611	614	620	629	[643]	(506)
250	306	346	362	382	406	439	461	464	492	493	[509]	456
300	267	302	305	336	347	351	376	387	392	409	[417]	397
350	239	265	273	296	297	303	303	335	354	358	352	344
400	212	241	254	261	263	269	274	304	322	324	303	303
450	199	231	234	241	243	249	250	266	281	284	271	275
500	195	227	226	224	230	232	234	242	240	270	243	251
600	177	177	200	199	199	210	203	207	206	216	212	215
700	152	167	177	178	187	185	184	184	191	193	187	188
800	151	154	159	166	167	167	170	173	172	174	172	164
900	148	149	149	149	149	160	157	159	159	157	165	145
1000	138	142	142	145	145	147	147	146	146	150	157	—

VARIATION OF THE COEFFICIENT OF EXPANSION WITH PRESSURE

An inspection of Table 21 shows that at the ontset the coefficient of expansion changes with pressure, in the manner already specified by Regnault for pressures of a few atmospheres. It then passes through a maximum, which occurs at a pressure regularly increasing with temperature. The maxima in each vertical column of the tables are put in parentheses. During my first researches on this subject I believed that these maxima occur at the same pressure for which the product PV is a minimum ; but the more extended data of the present memoir show this law to be only approximate.

At the critical temperature the maximum coefficient of expansion evidently coincides with the critical pressure, since the former is then infinite. Under other conditions, depending on the form of the isotherm, this pressure is much smaller than the pressure corresponding to the minimum ordinate. The locus of maximum coefficients of expansion thus starts out from the critical point (in my first memoir the initial isotherm was taken at 35.1° C.); and since this is a point of double inflection, the inquiry is pertinent whether the locus in question is not identical with the point of inflection of the isotherms. This is not the case. For increasing temperature, the maximum coefficient is always encountered a little earlier than the minimum ordinate. It is thus comprehended between the locus of the summits of these ordinates and the locus of the points of inflection. Little by little it approaches the former, and ends by intersecting it in the region of its minimum abscissa.

Table 22 for ethylene leads to a series of results of an analogous character throughout.

The following table (23) shows for oxygen, hydrogen, nitrogen, and air that the diminution of the coefficient of expansion continues regularly even as far as 3000 atmospheres. The same fact is exhibited in Table 24 for the same gases throughout higher temperatures.

G

TABLE 23.—VALUES OF $\dfrac{1}{v}\dfrac{\Delta v}{\Delta t}$

P	OXYGEN	HYDROGEN	
Atmospheres	0°—15.6°	0°—15.4°	0°—47.3°
1000	.00236	.00200	.00206
1500	189	178	173
2000	164	152	152
2500	147	138	137
3000	134	128	129

P	NITROGEN		AIR	
Atmospheres	0°—16.0°	0°—46.6°	0°—15.7°	0°—45.1°
1000	.00193	.00191	.00206	.00197
1500	140	151	144	148
2000	133	131	116	126
2500	111	108	107	112
3000	098	098	110	105

VARIATION OF THE COEFFICIENT OF EXPANSION WITH
TEMPERATURE

The above tables, 21 and 22, show that the coefficient at first increases with temperature, passes a maximum, and thereafter diminishes.

Under constant pressures of successively increasing value the maximum occurs at temperatures which continually increase, while the maximum itself becomes less accentuated and finally vanishes within the limits of the tables. An increase of value alone remains, which, in its turn, gradually becomes less appreciable. At 1000 atmospheres the maximum is certainly still encountered at sufficiently high temperatures. To verify these observations it suffices to treat such gases as are much farther removed from their critical points. It is then manifest that the maximum has been reached or even passed in all cases, as, for instance, Table 24 fully evidences. All the coefficients are here notably smaller between 100° and 200° than between 0° and 100°. Hence the maximum has been passed.

$$\text{TABLE } 24.\text{—VALUES OF } \frac{1}{v}\frac{\Delta v}{\Delta t} = a$$

P	OXYGEN		HYDROGEN		NITROGEN		AIR	
	0° to 99.5°	99.5° to 199.5°	0° to 99.25°	99.25° to 209.2°	0° to 99.45°	99.45° to 199.50°	0° to 99.4°	99.4° to 200.4°
Atm.	a	a	a	a	a	a	a	a
100	.00486	—	—	—	—	—	.00444	—
200	534	.00300	.00332	.00242	.00433	.00280	455	.00287
300	512	297	314	231	402	267	422	275
400	459	280	295	221	358	250	371	261
500	405	264	278	214	315	235	331	241
600	357	245	261	204	282	219	294	222
700	320	226	249	196	256	204	269	207
800	288	212	237	189	236	189	244	194
900	261	198	226	182	218	179	226	182
1000	241	—	218	--	—	--	214	171

In Tables 21 and 22 the maxima relative to temperature were put in square brackets. It is seen that they make up a group which lies very close to the maxima relative to pressure, and which, like the latter, would appear with greater regularity if the limits of the temperature and the pressure intervals were both narrower. The real maxima coincide, as it were, accidentally with the numbers in the tables.

The locus of the maxima relative to temperature starts from the critical point, as did the locus of the maxima relative to pressure. It approaches the locus of minimum ordinates more rapidly than the latter, and intersects it sooner, as it were.

Below the critical temperature the first coefficients of each horizontal row in Table 21 refer to the liquid state, since the pressure exceeds the maximum vapor tension. At pressures lower than the critical pressure the coefficients for the gaseous state, properly so called, at once decrease. As early as 1870 I showed * that for the cases of carbon dioxide and sulphur dioxide the coefficients decrease regularly from 0° C. to above 300° under atmospheric pressure.

For pressures of a value higher than the critical pressure there is no further occasion to consider the distinction between the two coefficients which I have just explained; for the discontinuity no longer occurs under constant pressure. To obviate all misapprehension I have marked three coefficients with an asterisk (*), which, although corresponding to temperatures

* *Comptes Rendus*, July 4, 1870; *Annales de Chimie et de Physique*, 1872.

below the critical temperature, belong to the gaseous state (at 50 and 60 atmospheres).

To return to the gaseous state : It was shown above that in case of oxygen, hydrogen, nitrogen, and air the coefficient of expansion has passed beyond its maximum value even at ordinary temperatures. The following table (25*), containing values of $\frac{\Delta v}{\Delta t}$ not reduced to the unit of volume, proves that the coefficients are practically independent of temperature, oxygen alone excepted :

TABLE 25*.—VALUES OF $\frac{\Delta v}{\Delta t} \times 10^{8}$

P	OXYGEN		HYDROGEN		NITROGEN		AIR	
	0° to 99 5°	99.5° to 199.5°	0° to 99.25°	99.25° to 200.2°	0° to 99.5°	99.45° to 199.5°	0° to 99.4°	99.4° to 200°
Atm.								
100	4518	—	—	—	—	—	4320	—
200	2442	2095	1890	1835	2251	2086	2296	2095
300	1643	1440	1265	1222	1522	1414	1544	1422
400	1207	1072	947	919	1124	1066	1125	1084
500	937	856	754	741	877	859	887	859
600	756	703	624	615	718	715	720	706
700	634	592	535	526	609	609	615	602
800	541	512	468	460	530	525	533	525
900	470	450	415	409	469	463	469	463
1000	418	—	376	—	—	—	425	413

Thus the coefficients beginning with 0° C. are sensibly constant for a given pressure, the same fact which was brought out by Table 23 as far as the highest pressures. Hence the coefficients computed for the successive intervals vary nearly inversely as the successive initial volumes. This appears to be the law towards which the decreasing march which follows the maximum points converges for conditions of increasing temperature. This limiting state is reached sooner in proportion as the pressure is smaller.

Hence, at all pressures and sufficiently high temperatures, this simple law supervenes : the increment of volume is proportional to increment of temperature reproducing the case of perfect gases. In a general way only is the volume proportional to the absolute temperature increased by a constant ; for this constant diminishes as pressure decreases in such a way that if pressure is small enough, the law of the proportionality of volume and absolute temperature is encountered. This again is the law of sensibly perfect gases.

COEFFICIENTS OF EXPANSION AT CONSTANT VOLUME,

$$\beta = \frac{1}{p}\frac{dp}{dt}, \text{ AND PRESSURE COEFFICIENTS, } B = \frac{dp}{dt}$$

To avoid all confusion, I will call the values $\frac{\Delta p}{\Delta t}$ *pressure co-efficients.* This reserves for $\frac{1}{p}\frac{\Delta p}{\Delta t}$ the time-honored, but otherwise very curious designation of *expansion coefficient* at constant volume.

In Table 25, on page 102, computed by aid of the data in Tables 6, 12, 17, 18, these coefficients are given relative to the temperature interval inserted at the heads of the vertical columns, and for the constant volume in the first of these columns. The pressures indicated for carbon dioxide and ethylene under the caption "Initial Pressures" do not all refer to zero, but to the lower limit of temperature of the first mean coefficient on the same horizontal row. The tables contain a gap which corresponds to the region contained within the curve of liquefaction.

It must be borne in mind that reduction to the unit of pressure of the coefficients β has been accomplished by dividing by the pressure corresponding to the lower temperature of each interval — at variance, therefore, to the notation frequently adopted. I have already made a similar remark relative to the reduction of the coefficients of compressibility and of expansion under constant pressure to the unit of volume.

TABLE 25.—VALUES OF $B = \dfrac{\Delta p}{\Delta t}$ AND $\beta = \dfrac{1}{p}\dfrac{\Delta p}{\Delta t}$ AT CONSTANT VOLUME

CARBON DIOXIDE

CONSTANT VOLUMES $V\times10^5$	INITIAL PRESSURES ATMOSPH.	0°—20°		20°—40°		40°—60°		60°—80°		80°—100°		100°—137°		137°—198°		198°—258°	
		$B\times10^3$	$\beta\times10^4$	$B\times10^3$	$\beta\times10^5$	$B\times10^3$	$\beta\times10^6$	$B\times10^3$	$\beta\times10^6$	$B\times10^3$	$\beta\times10^6$	$B\times10^3$	$\beta\times10^6$	$B\times10^3$	$\beta\times10^6$	$B\times10^3$	$\beta\times10^4$
2385	31.0	200	645	200	543	190	487	190	444	195	419	176	302	180	316	175	257
1636	45.1	.	—	315	698	310	603	295	512	300	472	284	408	279	349	250	258
1300	51.1	—	—	430	841	405	678	395	583	395	522	376	450	369	378	333	277
1000	56.3	—	—	610	1083	580	845	550	686	550	602	519	507	525	432	458	299
768	76.6	—	—	—	—	810	1057	770	830	780	722	735	504	721	477	658	338
578	83.1	—	—	—	—	1155	1390	1130	1224	1125	874	1073	709	1082	566	983	380
428	87.8	—	—	—	—	1705	1940	1720	1411	1740	1113	1659	868	1697	672	1558	437
316	98.0	—	—	—	—	2790	2847	2885	1876	2975	1407	2838	1047	2926	778		
250	64.4	—	—	4530	7030	4775	3081	4775	1906	4875	1409	4743	1069	4833	781		
200	122.5	8875	7245	8525	2841	9025	1919	9075	1394	8275	994	—	—	—	—		
187	307.5	10625	3455	11525	2208	10150	1352										

ETHYLENE

CONSTANT VOLUMES $V\times10^5$	INITIAL PRESSURES ATMOSPH.	0°—20°		20°—40°		40°—60°		60°—80°		80°—100°		100°—137° *		137°—198° †			
		$B\times10^3$	$\beta\times10^4$	$B\times10^3$	$\beta\times10^5$	$B\times10^3$	$\beta\times10^6$	$B\times10^3$	$\beta\times10^6$	$B\times10^3$	$\beta\times10^6$	$B\times10^3$	$\beta\times10^4$	$B\times10^3$	$\beta\times10^6$		
1666.6	37.0	310	838	300	694	300	610	305	553	300	489	285	424	285	366		
1125.0	52.5	—	—	505	962	495	791	490	676	475	577	458	499	459	421		
833.3	57.9	—	—	735	1271	720	992	710	816	720	711	696	602	688	485		
642.8	61.3	—	—	1030	1680	1035	1264	1045	1085	1040	842	1018	705	1016	557		
500.0	65.4	—	—	1475	2255	1500	1579	1525	1220	1525	981	1528	821	1498	611		
416.6	74.4	—	—	2080	2796	2170	1871	2130	1375	2220	1099	2104	854	2118	651		
350.0	49.0	2075	6071	3125	2980	3175	1854	3175	1354	3310	1111	3048	837	3025	632		
300.0	124.5	4750	3815	4610	2100	4715	1513	4650	1145	4800	960	4413	742	—	—		
285.7	185.5	5145	2235	5533	1848	5200	1316	5875	1075	5475	903	—	—	—	—		

* 100°—137.5°.   † 137.5°—198.5°.

THE LAWS OF GASES

TABLE 26.—VALUES OF B AND b

PRESSURES AT ZERO.	CONSTANT VOLUMES	HYDROGEN				CONSTANT VOLUMES	OXYGEN			
		0°—99.2°		99.2°—200.5°			0°—99.5°		99.5°—199.5°	
ATM.	$V\times10^6$	$B\times10^3$	$\beta\times10^5$	$B\times10^3$	$\beta\times10^5$	$V\times10^6$	$B\times10^3$	$\beta\times10^5$	$B\times10^3$	$\beta\times10^6$
100	10690	378	373	366	267	9265	492	492	470	315
200	5690	766	383	742	269	4570	1226	613	1115	357
300	4030	1149	383	1129	272	3208	2090	696	1900	374
400	3207	1521	380	1475	268	2629	2924	731	2570	372
500	2713	1895	379	1842	267	2312	3698	740	—	—
600	2386	2256	376	—	—	—	—	—	—	—
700	2149	3710	371	—	—	—	—	—	—	—

PRESSURES AT ZERO.	CONSTANT VOLUMES	NITROGEN				CONSTANT VOLUMES	AIR			
		0°—99.4°		99.4°—199.6°			0°—99.4°		99.4°—200.4°	
ATM.	$V\times10^6$	$B\times10^3$	$\beta\times10^5$	$B\times10^3$	$\beta\times10^5$	$V\times10^6$	$B\times10^3$	$\beta\times10^5$	$B\times10^3$	$\beta\times10^5$
100	9910	462	462	460	315	9730	462	462	465	319
200	5195	1075	537	1070	349	5050	1105	552	1090	351
300	3786	1748	582	1700	359	3658	1800	600	1742	364
400	3142	2382	595	2320	364	3036	2470	617	2327	360
500	2780	2982	596	—	—	2680	3085	617	—	—
600	2543	3582	597	—	—	2450	3718	620	—	—

TABLE 27.—VALUES OF B AND β

PRESSURES AT ZERO.	CONSTANT VOLUMES	HYDROGEN				CONSTANT VOLUMES	OXYGEN			
		.6°—15.4°		0°—47.3°				0°—15.6°		
ATM.	$V\times10^6$	$B\times10^3$	$\beta\times10^5$	$B\times10^3$	$\beta\times10^5$	$V\times10^6$		$B\times10^3$	$\beta\times10^5$	
600	—	—	—	—	—	2117.0	—	4423	737	
800	—	—	—	—	—	1880.0	—	5641	705	
1000	1725.0	3571	357	3467	347	1736.0	—	6795	679	
1200	1557.5	4155	346	4017	335	1635.0	—	8013	668	
1500	1380.0	5129	342	5010	334	1497.5	(1600 atm)	10513	657	
1800	1258.0	5779	321	5729	318	—	—	—	—	
2000	1194.5	6428	321	6342	317	1408.0	—	12051	602	
2400	1097.5	7662	319	7315	305	1343.5	—	13974	580	
2800	1024.5	8117	325	—	—	1304.0	(2700 atm.)	14423	538	

PRESSURES AT ZERO.	CONSTANT VOLUMES	NITROGEN				CONSTANT VOLUMES	AIR			
		0°—16.0°		0°—43.6°			0°—15.7°		0°—45.1°	
ATM.	$V\times10^6$	$B\times10^3$	$\beta\times10^5$	$B\times10^3$	$\beta\times10^5$	$V\times10^6$	$B\times10^3$	$\beta\times10^5$	$B\times10^3$	$\beta\times10^5$
800	—	—	—	—	—	2171.0	4841	605	4591	574
1000	2070.0	5500	550	5481	548	1999.0	5668	567	5543	554
1200	1946.0	6125	510	6284	524	1883.0	6051	504	6075	506
1500	1813.5	7060	471	7155	477	1754.0	7006	467	7273	485
1800	1714.5	8562	475	8440	469	1662.0	7900	439	8115	451
2000	1663.5	9375	468	9197	460	1613.0	8344	417	8736	437
2400	1583.5	10750	448	10505	437	1533.5	9681	403	9888	412
2800	1525.0	11875	424	—	—					

VARIATION OF THE COEFFICIENTS B AND β WITH VOLUME

The pressure coefficient B is seen to increase very rapidly when volume decreases—*i.e.*, when the initial pressure at zero increases. The coefficient β (Table 25) at first increases for increasing volume, and thereafter passes through a maximum which is much less pronounced when the temperature is higher. Finally β decreases with increasing volume. In case of nitrogen the maximum is not yet reached between 0° and 200° within the pressure limits given in Table 26. For hydrogen, on the contrary, its occurrence falls within the same limits, since β is then sensibly constant. For the highest pressures the maximum has been passed by in case of each of the four gases in Table 27.

VARIATION OF THE COEFFICIENTS B AND β WITH TEMPERATURE

Generally speaking, the coefficient B varies very little with temperature. An inspection of the table shows for carbon dioxide between 0° and 100° that this variation is quite insignificant. This is the identical result reached in my research* of 1881. Some time after Messrs. W. Ramsay and Sidney Young published important researches on the same subject, to which I shall recur on another occasion.

Between 100° and 260°, B shows a slight diminution. This is also true for the case of ethylene.

For hydrogen, air, and nitrogen the variation of B between 0° and 200° (Table 26) is scarcely apparent, particularly after the pressures approach the high values. A similar influence may be drawn for the other three gases (Table 27) for a pressure interval quite up to the highest pressures. It must be observed, however, that these results are restricted to smaller temperature intervals.

It appears to follow from the results as a whole that the variation of the pressure coefficient with temperature, *always very small*, quite vanishes at sufficiently high temperatures and probably at all temperatures under sufficiently high pressures. This is evidenced by the results shown by those gases which, within the temperature limits of the present research, are al-

* *Annales de Chimie et de Physique*, 5ᵉ Série, vol. xxii.

ready in a thermal state far above their critical points. Under these conditions the pressures corresponding to constancy of volume are not proportional to the respective absolute temperatures; they are proportional to them when each is diminished by a constant function of volume only. This constant is numerically a number of degrees, and it at first increases rapidly when volume diminishes; thereafter it passes through a maximum, decreases passing through zero into negative values, and continues to decrease in absolute value. Whenever this constant vanishes the gas is clearly characterized by the law of perfect gases, and this happens in the case of hydrogen at about 800 atmospheres. It is exceedingly remarkable that under these special conditions the value of the pressure coefficient is nearly equal to the value which holds for normal pressure—*i.e.*, to that attributed to gases when they approach as nearly as possible to the state of a perfect gas. It would be extremely interesting to discover whether the observation in question is of general significance. Unfortunately, the other gases studied have not under the highest pressures applied reached the state for which the constant in question vanishes.

The variations of the constant β may be deduced from what has just been stated. In every case this coefficient for a given volume varies very nearly inversely to pressure. In the region comprised within the curve of liquefaction, and corresponding to the gap in Table 25 (carbon dioxide), there is no true pressure coefficient. The values $\frac{dp}{dt}$ now refer to the maximum vapor tensions and no longer vary with volume. Necessarily an abrupt variation of these values occurs on breaking across the curve of liquefaction, excepting, perhaps, the line of equal volumes, which passes through the critical point with an inversion of the sign of the variation on one side or the other of this line. Indeed, it is easily observed that the values of $\frac{dp}{dt}$ for the lines of equal volume passing above the critical point and near the curve of saturation are smaller on the outside than on the inside of this curve. The contrary will be the case for the lines of equal volume which pass below the critical point. In every case the above inferences relative to pressure coefficients seem to be immediately applicable as soon as the curve of liquefaction is left behind.

105

The isotherms below the critical point are difficult to map out in those parts which are contiguous with the curve of liquefaction. This curve cannot be obtained by means of the above experiments as accurately as may be done by comparing the densities of the liquid and of the vapor obtained in experiments specially designed for this purpose. I have carried out measurements of this kind for carbon dioxide between zero and the critical point; but I will not enter into details relative to these results,* as they lie outside of the scope of the present investigation, beyond giving a tabulated view of the data. The agreement between the present values of maximum vapor tension and those contained in the above tables is apparent. The same research furnished the following elements of the critical point:

Critical temperature............... 31.35°
Critical pressure.................. 72.9 *atm.*
Critical density.................. 0.464

TABLE 28.—DATA FOR CARBON DIOXIDE

T	DENSITY OF THE LIQUID	DENSITY OF THE VAPOR	MAXIMUM VAPOR TENSION	*T*	DENSITY OF THE LIQUID	DENSITY OF THE VAPOR	MAXIMUM VAPOR TENSION
Deg.			*Atm.*	*Deg.*			*Atm.*
0	.914	.096	34.3	18	.786	.176	53.8
1	.910	.099	35.2	19	.776	.183	55.0
2	.906	.103	36.1	20	.766	.190	56.3
3	.900	.106	37.0	21	.755	.199	57.6
4	.894	.110	38.0	22	.743	.208	59.0
5	.888	.114	39.0	23	.731	.217	60.4
6	.882	.117	40.0	24	.717	.228	61.8
7	.876	.121	41.0	25	.703	.240	63.3
8	.869	.125	42.0	26	.688	.252	64 7
9	.863	.129	43.1	27	.671	.266	66.2
10	.856	.133	44.2	28	.653	.282	67.7
11	.848	.137	45.3	29	.630	.303	69.2
12	.841	.142	46.4	30	.598	.334	70.7
13	.831	.147	47.5	30.5	.574	.356	71.5
14	.822	.152	48.7	31.0	.536	.392	72.3
15	.814	.158	50.0	31.25	.497	.422	72.8
16	.806	.164	51.2	31.35	.464	.464	72.9
17	.796	.170	52.4				

* *Comptes Rendus*, May 16, 1892; June 7, 1892. Cf. *Journal de Physique*, 1892; *Séances de la Soc. de Physique*, 1892.

I have not up to the present time been able to repeat the same work for ethylene.

Certain other properties of the isotherms for carbon dioxide and ethylene, as well as divers inquiries of a more theoretical kind, are not in place here. In the present memoir, as well as in the following work relating to liquids, I have purposed merely to exhibit the experimental methods, to publish the numerical results obtained, and to deduce from them such general laws as result from inspection.

ÉMILE HILAIRE AMAGAT was born on the 2d of January, 1841, at *St. Satur*, a village in the *arrondissement de Sancerre*, in the *Département du Cher*. It was at first his intention to be a technical chemist, but he abandoned this career almost at the very outset in preference of one in pure science.

For several years Amagat was *préparateur* of the celebrated Berthelot at the *Collège de France*. After this (between 1867 and 1872) he was called to Switzerland, where he served as professor at the *Lycée de Fribourg*. It was there that Amagat completed his *thèse de doctorat*, being formally honored with this degree in Paris in 1872.

Returning to France, he was successively made professor at the *Lycée d'Alençon*, *à l'École Normale Spéciale de Cluny*, and in 1877 was appointed professor of physics in the *Faculté Libre des Sciences* of Lyons. In this institution, then merging into active existence, he created the department of physics, and in it conducted his most famous researches.

He left Lyons in 1891 for Paris, where he resides at present in the official position of *examinateur à l'École Polytechnique*.

Amagat has been correspondent of the *Institut de France* (Académie des Sciences, Section de Physique) since 1889. He was elected a foreign member of the *Royal Society of London* in 1897, and of the *Royal Society of Edinburgh* in the same year. He is honorary member of the *Société Hollandaise des Sciences*, of the *Société Scientifique de Bruxelles*, of the *Philosophical Society of Manchester*, etc., etc.

BIBLIOGRAPHY

AMONG Amagat's papers those bearing particularly on the laws of gases may be summarized for the reader's convenience as follows. A complete list of Amagat's researches will be found in a pamphlet published by Gautier-Villars et fils, Paris, 1896, entitled : *Notice sur les Travaux Scientifiques de M. E. H. Amagat.*

De l'influence de la température sur les écartes de la loi de Mariotte ; *C. R.,* lxviii., p. 1170, 1869.

Sur la compressibilité du gaz ; *C. R.,* lxxi., p. 67, 1870.

Sur la dilatation et la compress. des gaz ; *C. R.,* lxxiii., p. 183, 1871.

Sur la compress. de l'hydrogène et de l'air à des températures élevées ; *C. R.,* lxxv., p. 479, 1872.

Sur la dilatation des gaz humides ; *C. R.,* lxxiv., p. 1299, 1872.

Compress. de l'air et de l'hydrogène à des températures élevées ; *Annales de Chimie et de Physique* (4), xxviii., 1873.

Dilat. et compress. des gaz à divers températures ; *Annales de Chimie et de Physique* (4), xxix., 1873.

Recherches sur l'élasticité de l'air sur de faibles pressions ; *C. R.,* lxxxii., p. 914, 1876 ; ibid., *Annales de Chimie et de Physique* (5), viii., 1876.

Sur la compress des gaz à dépressions élevées ; *C. R.,* lxxxvii., p. 432, 1878. (Preliminary work at Fort Saint-Just.)

Expériences du puits Verpilleux ; *C. R.,* lxxxviii., p. 336, 1879.

Sur la compress. des divers gaz à des pressions élevées ; *C. R.,* lxxxix., p. 439, 1879.

Influence de la température sur la compress. des gaz sous de fortes pressions; *C. R.,* xc., p. 994, 1880.

Sur la dilatation et la compress. des gaz sous de fortes pressions ; *C. R.,* xci., p. 428, 1880.

Sur la compress. de l'oxygène, et l'action de ce gaz sur le mercure, etc.; *C. R.,* xci., p. 812, 1880.

Sur la compress. de l'acide carbonique et de l'air sous faible pression et temp. élevée ; *C. R.,* xciii., p. 306, 1881.

Mémoire sur la compressibilité des gaz aux fortes pressions ; *Annales de Chimie et de Physique* (5), xxii., 1881.

Sur la relation ϕ (p, v, t)=0 relative aux gaz, etc.; *C. R.,* xciv., p. 847, 1882.

Sur l'élasticité des gaz raréfiés ; *C. R.,* xcv., p. 281, 1882.

Sur .... la compress. du gaz azote ; *C. R.,* xcv., p. 638, 1882.

MEMOIRS ON THE LAWS OF GASES

Sujets relatifs à l'étude du gaz ; *Annales Chimie et de Physique* (5), xxviii., 1883.

Mémoire sur la compress. de l'air et de l'acide carbonique . . ; *Annales de Chimie et de Physique* (5), xxviii., 1883.

Mémoire sur la compress. de l'air, de l'hydrogène et de l'acide carbonique raréfiés ; *Annales de Chimie et de Physique* (5), xxviii., 1883.

Sur une forme nouvelle de la fonction $\phi\ (p, v, t)=0$; *Annales de Chimie et de Physique* (5), xxviii., 1883.

Résultats pour servir aux calculs des manomètres à gaz ; *C. R.*, xcix., p. 1017, 1884.

Note relative à une erreur . . . . ; *C. R.*, xcix., p. 1153, 1884.

Sur la densité limite et de volume atomique des gaz, etc.; *C. R.*, c., p. 633, 1885.

Sur la volume atomique de l'oxygène; *C. R.*, cii., p. 1100, 1886.

Compressibilité des gaz : oxygène, hydrogène, azote et air jusqu'à 3000 atm.; *C. R.*, cvii., p. 522, 1888.

Nouvelle méthode pour l'étude de la compress. et de la dilatation des liquides et des gaz . . . . ; *C. R.*, cxi., p. 871, 1890.

Sur la détermination de la densité des gaz et de leur vapeur saturée . . . . ; *C. R.*, cxiv., p. 1093, 1892 ; ibid., *C. R.*, cxiv., p. 1322, 1892 ; *Journal de Physique*, p. 288, 1892.

Sur les lois de dilatations des gaz sous pression constante ; *C. R.*, cxv., p. 771, 1892.

Sur la comparaison des lois de dilatation des liquides et de celles des gaz, etc.; *C. R.*, cxv., p. 919, 1892.

Sur les lois de dilatation à volume constant des fluides . . . . ; *C. R.*, cxv., p. 1041, 1892.

Mémoires sur l'élasticité et la dilatation des fluides jusqu'aux très hautes pressions ; *Annales de Chimie et de Physique* (6), xxix., 1893.

Sur la pression intérieure dans le gaz ; *C. R.*, cxviii., p. 326, 1894 ; ibid., *C. R.*, cxviii., 566, 1894.

Sur la pression intérieure et le viriel . . . . ; *C. R.*, cxx., p. 489, 1895.

Vérification d'ensemble de la loi des états correspondants de Van der Waals; *C. R.*, cxxiii., p. 30, 1896 ; ibid., *C. R.*, cxxiii., p. 83, 1896.

The titles of a few relevant papers by other investigators follow:

RELATIONS BETWEEN PRESSURE, VOLUME, AND TEMPERATURE

Ramsay and Young. *Philosophical Transactions*, **177, 178, 180, 183**.

Barus, C., *Philosophical Magazine* (5) **30**, 338–361, 1890.·

Tait, P. G., "Challenger Reports," *Physics and Chemistry*, vol. ii., part iv.; *Proceedings of the Royal Society of Edinburgh*, vols. xii., xiii., xx.

Leduc, A., *Journal de Physique* (3), **7**, 1898. *Annales de Chimie et de Physique*, **15**, 5–115, 1898.

Wroblewski, Wiedemann, *Annalen*, **29**, 428, 1886.

Rose-Innes, *Philosophical Magazine*, **44, 45**, 1897, 1898.

MEMOIRS ON THE LAWS OF GASES

CONTINUITY OF LIQUID AND GASEOUS STATES

Van der Waals,   "On the Continuity of the Liquid and Gaseous States."
        Translation.   London, 1890.   (Original Dutch edition,
        1873.)
Clausius, R.,   Wiedemann, *Annalen*, **9**, 337, 1880.
Sarrau,   *Comptes Rendus*, **110**, 880, 1890.
Ramsay and Young, *Philosophical Magazine* (5), **23**, **24**, 1887.
Brillouin, M.,   *Journal de Physique* (3), **2**, 113, 1893.
Tait, P. G.,   *Transactions of the Royal Society of Edinburgh*, **36**, 1891.
        Nature, **44**, **45**, 1891.
Rayleigh, Lord,   *Nature*, **44**, **45**, 1891.
Bakker, G.,   *Zeitschrift für Physikalische Chemie*, **21**, 127, 1896.
Young, S.,   *Philosophical Magazine* (5), **33**, 153, 1892 (**37**, 1, 1894).

CRITICAL STATE

Andrews,   *Philosophical Transactions*, **166**, 421–449, 1876.
Cailletet and Colardeau, *Annales de Chimie et de Physique* (6), **25**, 519, 1892.
Mathias, E.,   *Journal de Physique* (3), **1**, 53, 1892.
Kuenen, J. P.,   *Philosophical Magazine*, **44**, 1897.

THE END.

TEXT-BOOKS IN PHYSICS

THEORY OF PHYSICS

By JOSEPH S. AMES, Ph.D., Associate Professor of Physics in Johns Hopkins University. Crown 8vo, Cloth, $1 60; by mail, $1 75.

In writing this book it has been the author's aim to give a concise statement of the experimental facts on which the science of physics is based, and to present with these statements the accepted theories which correlate or "explain" them. The book is designed for those students who have had no previous training in physics, or at least only an elementary course, and is adapted to junior classes in colleges or technical schools. The entire subject, as presented in the work, may be easily studied in a course lasting for the academic year of nine months.

Perhaps the best general introduction to physics ever printed in the English language. . . . A model of comprehensiveness, directness, arrangement, and clearness of expression. . . . The treatment of each subject is wonderfully up to date for a text-book, and does credit to the system which keeps Johns Hopkins abreast of the times. Merely as an example of lucid expression and of systematization the book is worthy of careful reading.—*N. Y. Press.*

Seems to me to be thoroughly scientific in its treatment and to give the student what is conspicuously absent in certain well-known text-books on the subject—an excellent perspective of the very extensive phenomena of physics. — PROFESSOR F. E. BEACH, *Sheffield Scientific School of Yale University.*

A MANUAL OF EXPERIMENTS IN PHYSICS

Laboratory Instruction for College Classes. By JOSEPH S. AMES, Ph.D., Associate Professor of Physics in Johns Hopkins University, author of "Theory of Physics," and WILLIAM J. A. BLISS, Associate in Physics in Johns Hopkins University. 8vo, Cloth, $1 80; by mail, $1 95.

I have examined the book, and am greatly pleased with it. It is clear and well arranged, and has the best and newest methods. I can cheerfully recommend it as a most excellent work of its kind.—H. W. HARDING, *Professor Emeritus of Physics, Lehigh University.*

I think the work will materially aid laboratory instructors, lead to more scientific training of the students, and assist markedly in incentives to more advanced and original research.—LUCIEN I. BLAKE, *Professor of Physics, University of Kansas.*

It is written with that clearness and precision which are characteristic of its authors. I am confident that the book will be of great service to teachers and students in the physical laboratory.—HARRY C. JONES, Ph.D., *Instructor in Physical Chemistry, Johns Hopkins University.*

NEW YORK AND LONDON
HARPER & BROTHERS, PUBLISHERS

STANDARDS IN NATURAL SCIENCE

COMPARATIVE ZOOLOGY

Structural and Systematic. For use in Schools and Colleges. By JAMES ORTON, Ph.D. New edition, revised by CHARLES WRIGHT DODGE, M.S., Professor of Biology in the University of Rochester. With 350 illustrations. Crown 8vo, Cloth, $1 80 ; by mail, $1 96.

The distinctive character of this work consists in the treatment of the whole Animal Kingdom as a unit ; in the comparative study of the development and variations of organs and their functions, from the simplest to the most complex state ; in withholding Systematic Zoology until the student has mastered those structural affinities upon which true classification is founded ; and in being fitted for High Schools and Mixed Schools by its language and illustrations, yet going far enough to constitute a complete grammar of the science for the undergraduate course of any college.

INTRODUCTION TO ELEMENTARY PRACTICAL BIOLOGY

A Laboratory Guide for High Schools and College Students. By CHARLES WRIGHT DODGE, M.S., Professor of Biology, University of Rochester. Crown 8vo, Cloth, $1 80 ; by mail, $1 95.

Professor Dodge's manual consists essentially of questions on the structure and the physiology of a series of common animals and plants typical of their kind—questions which can be answered only by actual examination of the specimen or by experiment. Directions are given for the collection of specimens, for their preservation, and for preparing them for examination ; also for performing simple physiological experiments. Particular species are not required, as the questions usually apply well to several related forms.

THE STUDENTS' LYELL

A Manual of Elementary Geology. Edited by JOHN W. JUDD, C.B., LL.D., F.R.S., Professor of Geology, and Dean of the Royal College of Science, London. With a Geological Map, and 736 Illustrations in the Text. New, revised edition. Crown 8vo, Cloth, $2 25 ; by mail, $2 39.

The progress of geological science during the last quarter of a century has rendered necessary very considerable additions and corrections, and the rewriting of large portions of the book, but I have everywhere striven to preserve the author's plan and to follow the methods which characterize the original work.—*Extract from the Preface of the Revised Edition.*

NEW YORK AND LONDON

HARPER & BROTHERS, PUBLISHERS